FEYNMAN'S THESIS
A New Approach to Quantum Theory

FEYNMAN'S THESIS

A New Approach to Quantum Theory

Editor

Laurie M. Brown

Northwestern University, USA

NEW JERSEY • LONDON • SINGAPORE • BEIJING • SHANGHAI • HONG KONG • TAIPEI • CHENNAI

Published by

World Scientific Publishing Co. Pte. Ltd.
5 Toh Tuck Link, Singapore 596224
USA office: 27 Warren Street, Suite 401-402, Hackensack, NJ 07601
UK office: 57 Shelton Street, Covent Garden, London WC2H 9HE

Cover image: AIP Emilio Segrè Visual Archives, Physics Today Collection

ISBN 981-256-366-0

Printed in USA.

Contents

Preface

Since Richard Feynman's death in 1988 it has become increasingly evident that he was one of the most brilliant and original theoretical physicists of the twentieth century.[1] The Nobel Prize in Physics for 1965, shared with Julian Schwinger and Sin-itiro Tomonaga, rewarded their independent path-breaking work on the renormalization theory of quantum electrodynamics (QED). Feynman based his own formulation of a consistent QED, free of meaningless infinities, upon the work in his doctoral thesis of 1942 at Princeton University, which is published here for the first time. His new approach to quantum theory made use of the Principle of Least Action and led to methods for the very accurate calculation of quantum electromagnetic processes, as amply confirmed by experiment. These methods rely on the famous "Feynman diagrams," derived originally from the path integrals, which fill the pages of many articles and textbooks. Applied first to QED, the diagrams and the renormalization procedure based upon them also play a major role in other quantum field theories, including quantum gravity and the current "Standard Model" of elementary particle physics. The latter theory involves quarks and leptons interacting through the exchange of renormalizable Yang–Mills non-Abelian gauge fields (the electroweak and color gluon fields).

The path-integral and diagrammatic methods of Feynman are important general techniques of mathematical physics that have many applications other than quantum field theories: atomic and molecular scattering, condensed matter physics, statistical mechanics, quantum liquids and solids, Brownian motion, noise, etc.[2] In addition to

[1] Hans Bethe's obituary of Feynman [*Nature* **332** (1988), p. 588] begins: "Richard P. Feynman was one of the greatest physicists since the Second World War and, I believe, the most original."

[2] Some of these topics are treated in R. P. Feynman and A. R. Hibbs, *Quantum Mechanics and Path Integrals* (McGraw-Hill, Massachusetts, 1965). Also see M. C. Gutzwiller, "Resource Letter ICQM-1: The Interplay Between Classical and Quantum Mechanics," *Am. J. Phys.* **66** (1998), pp. 304–24; items 71–73 and 158–168 deal with path integrals.

its usefulness in these diverse fields of physics, the path-integral approach brings a new fundamental understanding of quantum theory. Dirac, in his transformation theory, demonstrated the complementarity of two seemingly different formulations: the matrix mechanics of Heisenberg, Born, and Jordan and the wave mechanics of de Broglie and Schrödinger. Feynman's independent path-integral theory sheds new light on Dirac's operators and Schrödinger's wave functions, and inspires some novel approaches to the still somewhat mysterious interpretation of quantum theory. Feynman liked to emphasize the value of approaching old problems in a new way, even if there were to be no immediate practical benefit.

Early Ideas on Electromagnetic Fields

Growing up and educated in New York City, where he was born on 11 May 1918, Feynman did his undergraduate studies at the Massachusetts Institute of Technology (MIT), graduating in 1939. Although an exceptional student with recognized mathematical prowess, he was not a prodigy like Julian Schwinger, his fellow New Yorker born the same year, who received his PhD in Physics from Columbia University in 1939 and had already published fifteen articles. Feynman had two publications at MIT, including his undergraduate thesis with John C. Slater on "Forces and Stresses in Molecules." In that work he proved a very important theorem in molecular and solid-state physics, which is now known as the Hellmann–Feynman theorem.[3]

While still an undergraduate at MIT, as he related in his Nobel address, Feynman devoted much thought to electromagnetic interactions, especially the self-interaction of a charge with its own field, which predicted that a pointlike electron would have an infinite mass. This unfortunate result could be avoided in classical physics, either by not calculating the mass, or by giving the theoretical electron an

[3] L. M. Brown (ed.), *Selected Papers of Richard Feynman, with Commentary* (World Scientific, Singapore, 2000), p. 3. This volume (hereafter referred to as *SP*) includes a complete bibliography of Feynman's work.

extended structure; the latter choice makes for some difficulties in relativistic physics.

Neither of these solutions are possible in QED, however, because the extended electron gives rise to non-local interaction and the infinite pointlike mass inevitably contaminates other effects, such as atomic energy level differences, when calculated to high accuracy. While at MIT, Feynman thought that he had found a simple solution to this problem: Why not assume that the electron does not experience any interaction with its own electromagnetic field? When he began his graduate study at Princeton University, he carried this idea with him. He explained why in his Nobel Address:[4]

> *Well, it seemed to me quite evident that the idea that a particle acts on itself is not a necessary one — it is a sort of silly one, as a matter of fact. And so I suggested to myself that electrons cannot act on themselves; they can only act on other electrons. That means there is no field at all. There was a direct interaction between charges, albeit with a delay.*

A new classical electromagnetic field theory of that type would avoid such difficulties as the infinite self-energy of the point electron. The very useful notion of a field could be retained as an auxiliary concept, even if not thought to be a fundamental one. There was a chance also that if the new theory were quantized, it might eliminate the fatal problems of the then current QED. However, Feynman soon learned that there was a great obstacle to this delayed action-at-a-distance theory: namely, if a radiating electron, say in an atom or an antenna, were not acted upon at all by the field that it radiated, then it would not recoil, which would violate the conservation of energy. For that reason, some form of radiative reaction is necessary.

---

[4] *SP*, pp. 9–32, especially p. 10.

The Wheeler–Feynman Theory

Trying to work through this problem at Princeton, Feynman asked his future thesis adviser, the young Assistant Professor John Wheeler, for help. In particular, he asked whether it was possible to consider that two charges interact in such a way that the second charge, accelerated by absorbing the radiation emitted by the first charge, itself emits radiation that reacts upon the first. Wheeler pointed out that there would be such an effect but, delayed by the time required for light to pass between the two particles, it could not be the force of radiation reaction, which is instantaneous; also the force would be much too weak. What Feynman had suggested was not radiation reaction, but the reflection of light!

However, Wheeler did offer a possible way out of the difficulty. First, one could assume that radiation always takes place in a totally absorbing universe, like a room with the blinds drawn. Second, although the principle of causality states that all observable effects take place at a time later than the cause, Maxwell's equations for the electromagnetic field possess a radiative solution other than that normally adopted, which is delayed in time by the finite velocity of light. In addition, there is a solution whose effects are advanced in time by the same amount. A linear combination of retarded and advanced solutions can also be used, and Wheeler asked Feynman to investigate whether some suitable combination in an absorbing universe would provide the required observed instantaneous radiative reaction?

Feynman worked out Wheeler's suggestion and found that, indeed, a mixture of one-half advanced and one-half retarded interaction in an absorbing universe would exactly mimic the result of a radiative reaction due to the electron's own field emitting purely retarded radiation. The advanced part of the interaction would stimulate a response in the electrons of the absorber, and their effect at the source (summed over the whole absorber) would arrive at just the right time and in the right strength to give the required radiation reaction force, without assuming any direct interaction of the electron with its own radiation field. Furthermore, no apparent violation of the principle of causality arises from the use of advanced radiation. Wheeler and Feynman further explored this beautiful theory in articles published in the Reviews of Modern Physics (RMP) in 1945 and 1949.[5] In the first of these articles, no less than four different proofs are presented of the important result concerning the radiative reaction.

Quantizing the Wheeler–Feynman theory (Feynman's PhD thesis): *The Principle of Least Action in Quantum Mechanics*

Having an action-at-a-distance classical theory of electromagnetic interactions without fields, except as an auxiliary device, the question arises as to how to make a corresponding quantum theory. To treat a classical system of interacting particles, there are available analytic methods using generalized coordinates, developed by Hamilton and Lagrange, corresponding canonical transformations, and the principle of least action.[6] The original forms of quantum mechanics, due to Heisenberg, Schrödinger, and Dirac, made use of the Hamiltonian approach and its consequences, especially Poisson brackets. To quantize the electromagnetic field it was represented, by Fourier transformation, as a superposition of plane waves having transverse, longitudinal, and timelike polarizations. A given field was represented as mathematically equivalent to a collection of harmonic oscillators. A system of interacting particles was then described by a Hamiltonian function of three terms representing respectively the particles, the field, and their interaction. Quantization consisted of regarding these terms as Hamiltonian *operators*, the field's Hamiltonian describing a suitable infinite set of quantized harmonic oscillators. The combination of longitudinal and timelike oscillators

[5] *SP*, p. 35–59 and p. 60–68. The second paper was actually written by Wheeler, based upon the joint work of both authors. It is remarked in these papers that H. Tetrode, W. Ritz, and G. N. Lewis had independently anticipated the absorber idea.

[6] W. Yourgrau and S. Mandelstam give an excellent analytic historical account in *Variational Principles in Dynamics and Quantum Theory* (Saunders, Philadelphia, 3rd edn., 1968).

was shown to provide the (instantaneous) Coulomb interaction of the particles, while the transverse oscillators were equivalent to photons. This approach, as well as the more general approach adopted by Heisenberg and Pauli (1929), was based upon Bohr's correspondence principle.

However, no method based upon the Hamiltonian could be used for the Wheeler–Feynman theory, either classically or quantum mechanically. The principal reason was the use of half-advanced and half-retarded interaction. The Hamiltonian method describes and keeps track of the state of the system of particles and fields at a given time. In the new theory, there are no field variables, and every radiative process depends on contributions from the future as well as from the past! One is forced to view the entire process from start to finish. The only existing classical approach of this kind for particles makes use of the principle of least action, and Feynman's thesis project was to develop and generalize this approach so that it could be used to formulate the Wheeler–Feynman theory (a theory possessing an action, but without a Hamiltonian). If successful, he should then try to find a method to quantize the new theory.[7]

The Introduction to the Thesis

Presenting his motivation and giving the plan of the thesis, Feynman's introductory section laid out the principal features of the (not yet published) delayed electromagnetic action-at-a-distance theory as described above, including the postulate that "fundamental (microscopic) phenomena in nature are symmetrical with respect to the interchange of past and future." Feynman claimed: "This requires that the solution of Maxwell's equation[s] to be used in computing the interactions is to be half the retarded plus half the advanced solution of Lienard and Wiechert." Although it would appear to contradict causality, Feynman stated that the principles of

[7] For a related discussion, including Feynman's PhD thesis, see S. S. Schweber, *QED and the Men Who Made It: Dyson, Feynman, Schwinger, and Tomonaga* (Princeton University Press, Princeton, 1994), especially pp. 389–397.

the theory "do in fact lead to essential agreement with the results of the more usual form of electrodynamics, and at the same time permit a consistent description of point charges and lead to a unique law of radiative damping .... It is shown that these principles are equivalent to the equations of motion resulting from a principle of least action."

To explain the spontaneous decay of excited atoms and the existence of photons, both seemingly contradicted by this view, Feynman argued that "an atom alone in empty space would, in fact, *not* radiate ... and all of the apparent quantum properties of light and the existence of photons may be nothing more than the result of matter interacting with matter directly, and according to quantum mechanical laws."

Two important points conclude the introduction. First, although the Wheeler–Feynman theory clearly furnished its motivation: "It is to be emphasized ... that the work described here is complete in itself without regard to its application to electrodynamics ... [The] present paper is concerned with the problem of finding a quantum mechanical description applicable to systems which in their classical analogue are expressible by a principle of least action, and not necessarily by Hamiltonian equations of motion." The second point is this: "All of the analysis will apply to non-relativistic systems. The generalization to the relativistic case is not at present known."

Classical Dynamics Generalized

The second section of the thesis discusses the theory of functionals and functional derivatives, and it generalizes the principle of least action of classical dynamics. Applying this method to the particular example of particles interacting through the intermediary of classical harmonic oscillators (an analogue of the electromagnetic field), Feynman shows how the coordinates of the oscillators can be eliminated and how their role in the interaction is replaced by a direct delayed interaction of the particles. Before this elimination process, the system consisting of oscillators and particles possesses a Hamiltonian but afterward, when the particles have direct interaction, no

Hamiltonian formulation is possible. Nevertheless, the equations of motion can still be derived from the principle of least action. This demonstration sets the stage for a similar procedure to be carried out in the quantized theory developed in the third and final section of the thesis.

In classical dynamics, the action is given by

$$S = \int L(q(t), \dot{q}(t))dt \,,$$

where L is a function of the generalized coordinates $q(t)$ and the generalized velocities $\dot{q} = dq/dt$, the integral being taken between the initial and final times t_0 and t_1, for which the set of q's have assigned values. The action depends on the paths $q(t)$ taken by the particles, and thus it is a functional of those paths. The *principle of least* action states that for "small" variations of the paths, the end points being fixed, the action S is an extremum, in most cases a minimum. An equivalent statement is that the functional derivative of S is zero. In the usual treatment, this principle leads to the Lagrangian and Hamiltonian equations of motion.

Feynman illustrates how this principle can be extended to the case of a particle (perhaps an atom) interacting with itself through advanced and retarded waves, by means of a mirror. An interaction term of the form $k^2\dot{x}(t)\dot{x}(t+T)$ is added to the Lagrangian of the particle in the action integral, T being the time for light to reach the mirror and return to the particle. (As an approximation, the limits of integration of the action integral are taken as negative and positive infinity.) A simple calculation, setting the variation of the action equal to zero, leads to the equation of motion of the particle. This shows that the force on the particle at time t depends on the particle's motion at times t, $t-T$, and $t+T$. That leads Feynman to observe: "The equations of motion cannot be described directly in Hamiltonian form."

After this simple example, there is a section discussing the restrictions that are needed to guarantee the existence of the usual constants of motion, including the energy. The thesis then treats the more complicated case of particles interacting via intermediate oscillators. It is shown how to eliminate the oscillators and obtain direct delayed action-at-a-distance. Interestingly, by making a suitable choice of the action functional, one can obtain particles either with or without self-interaction.

While still working on formulating the classical Wheeler–Feynman theory, Feynman was already beginning to adopt the over-all space-time approach that characterizes the quantization carried out in the thesis and in so much of his subsequent work, as he explained in his Nobel Lecture:[8]

> *By this time I was becoming used to a physical point of view different from the more customary point of view. In the customary view, things are discussed as a function of time in very great detail. For example, you have the field at this moment, a different equation gives you the field at a later moment and so on; a method, which I shall call the Hamiltonian method, a time differential method. We have, instead [the action] a thing that describes the character of the path throughout all of space and time. The behavior of nature is determined by saying her whole space-time path has a certain character. For the action [with advanced and retarded terms] the equations are no longer at all easy to get back into Hamiltonian form. If you wish to use as variables only the coordinates of particles, then you can talk about the property of the paths — but the path of one particle at a given time is affected by the path of another at a different time .... Therefore, you need a lot of bookkeeping variables to keep track of what the particle did in the past. These are called field variables .... From the overall space-time point of view of the least action principle, the field disappears as nothing but bookkeeping variables insisted on by the Hamiltonian method.*

Of the many significant contribution to theoretical physics that Feynman made throughout his career, perhaps none will turn out to

---

8 "The development of the space-time view of quantum electrodynamics," *SP*, pp. 9–32, especially p. 16.

be of more lasting value than his reformulation of quantum mechanics, complementing those of Heisenberg, Schrödinger, and Dirac.[9] When extended to the relativistic domain and including the quantized electromagnetic field, it forms the basis of Feynman's version of QED, which is now the version of choice of theoretical physics, and which was seminal in the development of the gauge theories employed in the Standard Model of particle physics.[10]

Quantum Mechanics and the Principle of Least Action

The third and final section of the thesis, together with the RMP article of 1949, presents the new form of quantum mechanics.[11] In reply to a request for a copy of the thesis, Feynman said he had not an available copy, but instead sent a reprint of the RMP article, with this explanation of the difference:[12]

> *This article contains most of what was in the thesis. The thesis contained in addition a discussion of the relation between constants of motion such as energy and momentum and invariance properties of an action functional. Further there is a much more thorough discussion of the possible gen-*

---

[9] The action principle approach was later adopted also by Julian Schwinger. In discussing these formulations, Yourgrau and Mandelstam comment: "One cannot fail to observe that Feynman's principle in particular — and this is no hyperbole — expresses the laws of quantum mechanics in an exemplary neat and elegant manner, notwithstanding the fact that it employs somewhat unconventional mathematics. It can easily be related to Schwinger's principle, which utilizes mathematics of a more familiar nature. The theorem of Schwinger is, as it were, simply a translation of that of Feynman into differential notation." (Taken from Yourgrau and Mandelstam's book [footnote 6], p. 128.)

[10] Although it had initially motivated his approach to QED, Feynman found later that the quantized version of the Wheeler–Feynman theory (that is, QED without fields) could not account for the experimentally observed phenomenon known as vacuum polarization. Thus in a letter to Wheeler (on May 4, 1951) Feynman wrote: "I wish to deny the correctness of the assumption that electrons act only on other electrons .... So I think we guessed wrong in 1941. Do you agree?"

[11] R. P. Feynman, "Space-time approach to non-relativistic quantum mechanics," *Rev. Mod. Phys.* **20** (1948) pp. 367–387 included here as an appendix. Also in *SP*, pp. 177–197.

[12] Letter to J. G. Valatin, May 11, 1949.

eralization of quantum mechanics to apply to more general functionals than appears in the Review article. Finally the properties of a system interacting through intermediate harmonic oscillators is discussed in more detail.

The introductory part of this third section of the thesis refers to Dirac's classical treatise for the usual formulation of quantum mechanics.[13]

However, Feynman writes that for those classical systems, which have no Hamiltonian form "no satisfactory method of quantization has been given." Thus he intends to provide one, based on the principle of least action. He will show that this method satisfies two necessary criteria: First, in the limit that $\hbar$ approaches zero, the quantum mechanical equations derived approach the classical ones, including the extended ones considered earlier. Second, for a system whose classical analogue does possess a Hamiltonian, the results are completely equivalent to the usual quantum mechanics.

The next section, "The Lagrangian in Quantum Mechanics" has the same title as an article of Dirac, published in 1933.[14] Dirac presents there an alternative version to a quantum mechanics based on the classical Hamiltonian, which is a function of the coordinates q and the momenta p of the system. He remarks that the Lagrangian, a function of coordinates and velocities, is more fundamental because the action defined by it is a relativistic invariant, and also because it admits a principle of least action. Furthermore, it is "closely connected to the theory of contact transformations," which has an important quantum mechanical analogue, namely, the transformation matrix $(q_t|q_T)$. This matrix connects a representation with the variables q diagonal at time T with a representation having the q's diagonal at time t. In the article, Dirac writes that $(q_t|q_T)$

[13] P. A. M. Dirac, The Principles of Quantum Mechanics (Oxford University Press, Oxford, 2nd edn., 1935). Later editions contain very similar material regarding the fundamental aspects to which Feynman refers.

[14] P. A. M. Dirac, in *Physikalische Zeitschrift der Sowjetunion*, Band 3, Heft 1 (1933), included here as an appendix. In discussing this material, Feynman includes a lengthy quotation from Dirac's *Principles*, 2nd edn., pp. 124–126.

"corresponds to" the quantity $A(tT)$, defined as

$$A(tT) = \exp\left[i\int_T^t Ldt/\hbar\right].$$

A bit later on, he writes that $A(t_T)$ "is the classical analogue of $(q_t|q_T)$."

When Herbert Jehle, who was visiting Princeton in 1941, called Feynman's attention to Dirac's article, he realized at once that it gave a necessary clue, based upon the principle of least action that he could use to quantize classical systems that do not possess a Hamiltonian. Dirac's paper argues that the classical limit condition for $\hbar$ approaching zero is satisfied, and Feynman shows this explicitly in his thesis. The procedure is to divide the time interval $t - T$ into a large number of small elements and consider a succession of transformations from one time to the next:

$$(q_t|q_T) = \iint\cdots\int (q_t|q_m)dq_m(q_m|q_{m-1})dq_{m-1}\cdots(q_2|q_1)dq_1(q_1|q_T).$$

If the transformation function has a form like $A(tT)$, then the integrand is a rapidly oscillating function when $\hbar$ is small, and only those paths $(q_T, q_1, q_2, \ldots, q_t)$ give an appreciable contribution for which the phase of the exponential is stationary. In the limit, only those paths are allowed for which the action is a minimum; i.e., for which $\delta S = 0$, with

$$S = \int_T^t Ldt.$$

For a very small time interval ε, the transformation function takes the form

$$A(t, t+\varepsilon) = \exp iL\varepsilon/\hbar,$$

where $L = L((Q-q)/\varepsilon, Q)$, and we have let $q = q_t$ and $Q = q_{t+\varepsilon}$. Applying the transformation function to the wave function $\psi(q,t)$ to obtain $\psi(Q, t+\varepsilon)$ and expanding the resulting integral equation to first order in ε, Feynman obtains the Schrödinger equation. His derivation is valid for any Lagrangian containing at most quadratic terms in the velocities. In this way he demonstrates two important points: In the first place, the derivation shows that the usual results of quantum mechanics are obtained for systems possessing a classical Lagrangian from which a Hamiltonian can be derived. Second, he shows that Dirac's $A(tT)$ is not merely an analogue of $(q_t|q_T)$, but is *equal* to it, for a small time ε, up to a normalization factor. For a single coordinate, this factor is $N = \sqrt{2\pi i\varepsilon\hbar/m}$.

This method turns out to be an extraordinarily powerful way to obtain Feynman's path-integral formulation of quantum mechanics, upon which much of his subsequent thinking and production was based. Successive application of infinitesimal transformations provides a transformation of the wave function over a finite time interval, say from time T to time t. The Lagrangian in the exponent can be approximated to first order in ε, and

$$\psi(Q,T) \cong \iint \cdots \int \exp\left\{\frac{i}{\hbar}\sum_{i=0}^{m}\left[L\left(\frac{q_{i+1}-q_i}{t_{i+1}-t_i}, q_{i+1}\right)(t_{i+1}-t_i)\right]\right\}$$
$$\times\ \psi(q_0,t_0)\frac{\sqrt{g_0}dq_0\cdots\sqrt{g_m}dq_m}{N(t_1-t_0)\cdots N(T-t_m)},$$

is the result obtained by induction, where $Q = q_{m+1}, T = t_{m+1}$, and the N's are the normalization factors (one for each q) referred to above. In the limit where ε goes to zero, the right-hand side is equal to $\psi(Q,T)$. Feynman writes: "The sum in the exponential resembles $\int_{t_0}^{T} L(q,\dot{q})dt$ with the integral written as a Riemann sum. In a similar manner we can compute $\psi(q_0,t_0)$ in terms of the wave function at a later time ..."

A sequence of q's for each t_i will, in the limit, define a path of the system and each of the integrals is to be taken over the entire range available to each q_i. In other words, the multiple integral is taken over all possible paths. We note that each path is continuous but not, in general, differentiable.

Using the idea of path integrals as in the expression above for $\psi(Q,T)$, Feynman considers expressions at a given time t_0, such as $\langle f(q_0)\rangle = \langle\chi|f(q_0)|\Psi\rangle$, which represents a quantum mechanical matrix element if χ and Ψ are different state functions or an expectation value if they represent the same state (i.e., $\chi = \Psi^*$). Path integrals relate the wave function $\psi(q_0, t_0)$ to an earlier time and the wave function $\chi(q_0, t_0)$ to a later time, which are taken as the distant past and future, respectively. By writing $\langle f(q_0) \rangle$ at two times separated by ε and letting ε approach zero, Feynman shows how to calculate the time derivative of $\langle f(q_t) \rangle$.

The next section of the thesis uses the language of functionals $F(q_i)$, depending on the values of the q's at the sequence of times t_i, to derive the quantum Lagrangian equations of motion from the path integrals. It shows the relation of these equations to q-number equations, such as $pq - qp = \hbar/i$ and discusses the relation of the Lagrangian formulation to the Hamiltonian one for cases where the latter exists. For example, the well-known result is derived that $HF - FH = (\hbar/i)\dot{F}$.

As was the case in the discussion of the classical theory, Feynman extends the formalism to the case of a more general action functional, beginning with the simple example of "a particle in a potential $V(x)$ and which also interacts with itself in a mirror, with half advanced and half retarded waves." An immediate difficulty is that the corresponding Lagrangian function involves two times. As a consequence, the action integral over the finite interval between times T_1 and T_2 is meaningless, because "the action might depend on values of $x(t)$ outside of this range." One can avoid this difficulty by formally letting the interaction vanish at times after large positive T_2 and before large negative T_1. Then for times outside the range of integration the particles are effectively free, so that wave functions can be defined at the endpoints. With this assumption the earlier discussions concerning functionals, operators, etc., can be carried through with the more general action functional.

However, the question as to whether a wave function or other wave-function-like object exists with the generalized Lagrangian is not solved in the thesis (and perhaps has never been solved). Although Feynman shows that much of quantum mechanics can be solved in terms of expectation values and transition amplitudes, at the end it is far from clear that it is possible to drop the very useful notion of the wave function (and if it is possible, it is probably not desirable to do so). A number of the pages of the thesis that follow are concerned with the question of the wave function, with conservation of energy, and with the calculation of transition probability amplitudes, including the development of a perturbation theory.

We shall not discuss these issues here, but continue to the last part of the thesis, where the forced harmonic oscillator is calculated. Based upon the path-integral solution of that problem, particles interacting through an intermediate oscillator are introduced and eventually the oscillators (i.e. the "field variables") are completely eliminated. Enrico Fermi had introduced the method of representing the electromagnetic field as a collection of oscillators and had eliminated the oscillators of longitudinal and timelike polarization to give the instantaneous Coulomb potential, as Feynman points out.[15] That had been the original aim of the thesis, to eliminate *all of the oscillators* (and hence the field) in order to quantize the Wheeler–Feynman action-at-a-distance theory. It turns out, however, that the elimination of all the oscillators was also very valuable in field theory having purely retarded interaction, and led in fact to the overall space-time point of view, to path integrals, and eventually to Feynman diagrams and renormalization.

We will sketch very briefly how Feynman handled the forced oscillator, using the symbol S for the generalized action. He wrote

$$S = S_0 + \int dt \left\{ \frac{m\dot{x}^2}{2} - \frac{m\omega^2 x^2}{2} + \gamma(t)x \right\},$$

where S_0 is the action of the other particles of the system of which the oscillator $[x(t)]$ is a part, and $\gamma(t)x$ is the interaction of the oscillator with the particles that form the rest of the system. If $\gamma(t)$ is a simple function of time (for example $\cos\omega_1 t$) then it represents a given force applied to the oscillator. However, more generally we are dealing with an oscillator interacting with another quantum system and $\gamma(t)$ is a functional of the coordinates of that system. Since the action $S - S_0$ depends quadratically and linearly on $x(t)$, the path integrals

[15] Feynman mentions in this connection Fermi's influential article "Quantum theory of Radiation," *Rev. Mod. Phys.* **4** (1932) pp. 87–132. In this paper, the result is assumed to hold; it was proven earlier by Fermi in "Sopra l'elettrodynamica quantistica," *Rendiconti della R. Accademia Nazionali dei Lincei* **9** (1929) pp. 881–887.

over the paths of the oscillator can be performed when calculating the transition amplitude of the system from the initial time 0 to the final time T. With $x(0) = x$ and $x(T) = x'$, Feynman calls the function so obtained $G_\gamma(x, x'; T)$, obtaining finally the formula for the transition amplitude

$$\langle \chi_T | 1 | \psi_0 \rangle_S = \int \chi_T(Q_m, x) e^{\frac{i}{\hbar} S_0[\ldots Q_i \ldots]} G_\gamma(x, x'; T) \psi_0(Q_0, x') \times dx dx' \frac{\sqrt{g} dQ_m \cdots \sqrt{g} dQ_0}{N_m \cdots N_1},$$

where the Q's are the coordinates of the system other than the oscillator.

By using the last expression in the problem of particles interacting through an intermediate oscillator having $x(0) = \alpha$ and $x(T) = \beta$, Feynman shows that the expected value of a functional of the coordinates of the particles alone (such as a transition amplitude) can be obtained with a certain action that does not involve the oscillator coordinates, but only the constants α and β.[16] This eliminates the oscillator from the dynamics of the problem. Various other initial and/or final conditions on the oscillator are shown to lead to a similar result. A brief section labeled "Conclusion" completes the thesis.

Laurie M. Brown
April 2005

The editor (LMB) thanks to Professor David Kiang for his invaluable assistance in copy-editing the retyped manuscript and checking the equations.

---

[16] In the abstract at the end of the thesis this conclusion concerning the interaction of two systems is summarized as follows: "It is shown that in quantum mechanics, just as in classical mechanics, under certain circumstances the oscillator can be completely eliminated, its place being taken by a direct, but, in general, not instantaneous, interaction between the two systems."

THE PRINCIPLE OF LEAST ACTION IN QUANTUM MECHANICS

RICHARD P. FEYNMAN

Abstract

A generalization of quantum mechanics is given in which the central mathematical concept is the analogue of the action in classical mechanics. It is therefore applicable to mechanical systems whose equations of motion cannot be put into Hamiltonian form. It is only required that some form of least action principle be available.

It is shown that if the action is the time integral of a function of velocity and position (that is, if a Lagrangian exists), the generalization reduces to the usual form of quantum mechanics. In the classical limit, the quantum equations go over into the corresponding classical ones, with the same action function.

As a special problem, because of its application to electrodynamics, and because the results serve as a confirmation of the proposed generalization, the interaction of two systems through the agency of an intermediate harmonic oscillator is discussed in detail. It is shown that in quantum mechanics, just as in classical mechanics, under certain circumstances the oscillator can be completely eliminated, its place being taken by a direct, but, in general, not instantaneous, interaction between the two systems.

The work is non-relativistic throughout.

I. Introduction

Planck's discovery in 1900 of the quantum properties of light led to an enormously deeper understanding of the attributes and behaviour of matter, through the advent of the methods of quantum mechanics. When, however, these same methods are turned to the problem of light and the electromagnetic field great difficulties arise which have not been surmounted satisfactorily, so that Planck's observations still remain without a consistent fundamental interpretation.[1]

As is well known, the quantum electrodynamics that have been developed suffer from the difficulty that, taken literally, they predict infinite values for many experimental quantities which are obviously quite finite, such as for example, the shift in energy of spectral lines due to interaction of the atom and the field. The classical field theory of Maxwell and Lorentz serves as the jumping-off point for this quantum electrodynamics. The latter theory, however, does not take over the ideas of classical theory concerning the internal structure of the electron, which ideas are so necessary to the classical theory to attain finite values for such quantities as the inertia of an electron. The researches of Dirac into the quantum properties of the electron have been so successful in interpreting such properties as its spin and magnetic moment, and the existence of the positron, that is hard to believe that it should be necessary in addition to attribute internal structure to it.

It has become, therefore, increasingly more evident that before a satisfactory quantum electrodynamics can be developed it will be necessary to develop a classical theory capable of describing charges without internal structure. Many of these have now been developed, but we will concern ourselves in this thesis with the theory of action at a distance worked out in 1941 by J. A. Wheeler and the author.[2]

The new viewpoint pictures electrodynamic interaction as direct interaction at a distance between particles. The field then becomes a mathematical construction to aid in the solution of problems involving these interactions. The following principles are essential to the altered viewpoint:

(1) The acceleration of a point charge is due to the sum of its interactions with other charged particles. A charge does not act on itself.

---

[1] It is important to develop a satisfactory quantum electrodynamics also for another reason. At the present time theoretical physics is confronted with a number of fundamental unsolved problems dealing with the nucleus, the interactions of protons and neutrons, etc. In an attempt to tackle these, meson field theories have been set up in analogy to the electromagnetic field theory. But the analogy is unfortunately all too perfect; the infinite answers are all too prevalent and confusing.

[2] Not published. See, however, *Phys. Rev.* **59**, 683 (1941).

(2) The force of interaction which one charge exerts on another is calculated by means of the Lorentz force formula, $F = e[E + \frac{v}{c} \times H]$, in which the fields are the fields generated by the first charge according to Maxwell's equations.
(3) The fundamental (microscopic) phenomena in nature are symmetrical with respect to interchange of past and future. This requires that the solution of Maxwell's equation to be used in computing the interactions is to be half the retarded plus half the advanced solution of Lienard and Wiechert.

These principles, at first sight at such variance with elementary notions of causality, do in fact lead to essential agreement with the results of the more usual form of electrodynamics, and at the same time permit a consistent description of point charges and lead to a unique law of radiative damping. That this is the case has been shown in the work already referred to (see note 2). It is shown that these principles are equivalent to the equations of motion resulting from a principle of least action. The action function (due to Tetrode,[3] and, independently, to Fokker[4]) involves only the coordinates of the particles, no mention of fields being made. The field is therefore a derived concept, and cannot be pictured as analogous to the vibrations of some medium, with its own degrees of freedom (for example, the energy density is not necessarily positive.) Perhaps a word or two as to what aspects of this theory make it a reasonable basis for a quantum theory of light would not be amiss.

When one attempts to list those phenomena which seem to indicate that light is quantized, the first type of phenomenon which comes to mind are like the photoelectric effect or the Compton effect. One is however, struck by the fact that since these phenomena deal with the interaction of light and matter their explanation may lie in the quantum aspects of matter, rather than requiring photons of light. This supposition is aided by the fact that if one solves the

[3] H. Tetrode, *Zeits. f. Physik* **10**, 317 (1922).
[4] A. D. Fokker, *Zeits. f. Physik* **38**, 386 (1929); *Physica* **9**, 33 (1929); *Physica* **12**, 145 (1932).

problem of an atom being perturbed by a potential varying sinusoidally with the time, which would be the situation if matter were quantum mechanical and light classical, one finds indeed that it will in all probability eject an electron whose energy shows an increase of $h\nu$, where ν is the frequency of variation of the potential. In a similar way an electron perturbed by the potential of two beams of light of different frequencies and different directions will make transitions to a state in which its momentum and energy is changed by an amount just equal to that given by the formulas for the Compton effect, with one beam corresponding in direction and wavelength to the incoming photon and the other to the outgoing one. In fact, one may correctly calculate in this way the probabilities of absorption and induced emission of light by an atom.

When, however, we come to spontaneous emission and the mechanism of the production of light, we come much nearer to the real reason for the apparent necessity of photons. The fact that an atom emits spontaneously at all is impossible to explain by the simple picture given above. In empty space an atom emits light and yet there is no potential to perturb the systems and so force it to make a transition. The explanation of modern quantum mechanical electrodynamics is that the atom is perturbed by the zero-point fluctuations of the quantized radiation field.

It is here that the theory of action at a distance gives us a different viewpoint. It says that an atom alone in empty space would, in fact, *not* radiate. Radiation is a consequence of the interaction with other atoms (namely, those in the matter which absorbs the radiation). We are then led to the possibility that the spontaneous radiation of an atom in quantum mechanics also, may not be spontaneous at all, but induced by the interaction with other atoms, and that all of the apparent quantum properties of light and the existence of photons may be nothing more than the result of matter interacting with matter directly, and according to quantum mechanical laws.

An attempt to investigate this possibility and to find a quantum analogue of the theory of action at a distance, meets first the difficulty that it may not be correct to represent the field as a set of harmonic oscillators, each with its own degree of freedom, since the field in actuality is entirely determined by the particles. On the other hand, an attempt to deal quantum mechanically directly with the particles, which would seem to be the most satisfactory way to proceed, is faced with the circumstance that the equations of motion of the particles are expressed classically as a consequence of a principle of least action, and cannot, it appears, be expressed in Hamiltonian form.

For this reason a method of formulating a quantum analog of systems for which no Hamiltonian, but rather a principle of least action, exists has been worked out. It is a description of this method which constitutes this thesis. Although the method was worked out with the express purpose of applying it to the theory of action at a distance, it is in fact independent of that theory, and is complete in itself. Nevertheless most of the illustrative examples will be taken from problems which arise in the action at a distance electrodynamics. In particular, the problem of the equivalence in quantum mechanics of direct interaction and interaction through the agency of an intermediate harmonic oscillator will be discussed in detail. The solution of this problem is essential if one is going to be able to compare a theory which considers field oscillators as real mechanical and quantized systems, with a theory which considers the field as just a mathematical construction of classical electrodynamics required to simplify the discussion of the interactions between particles. On the other hand, no excuse need be given for including this problem, as its solution gives a very direct confirmation, which would otherwise be lacking, of the general utility and correctness of the proposed method of formulating the quantum analogue of systems with a least action principle.

The results of the application of these methods to quantum electrodynamics is not included in this thesis, but will be reserved for a future time when they shall have been more completely worked out.

It has been the purpose of this introduction to indicate the motivation for the problems which are discussed herein. It is to be emphasized again that the work described here is complete in itself without regard to its application to electrodynamics, and it is this circumstance which makes it appear advisable to publish these results as an independent paper. One should therefore take the viewpoint that the present paper is concerned with the problem of finding a quantum mechanical description applicable to systems which in their classical analogue are expressible by a principle of least action, and not necessarily by Hamiltonian equations of motion.

The thesis is divided into two main parts. The first deals with the properties of classical systems satisfying a principle of least action, while the second part contains the method of quantum mechanical description applicable to these systems. In the first part are also included some mathematical remarks about functionals. All of the analysis will apply to non-relativistic systems. The generalization to the relativistic case is not at present known.

II. Least Action in Classical Mechanics

1. The Concept of a Functional

The mathematical concept of a functional will play a rather predominant role in what is to follow so that it seems advisable to begin at once by describing a few of the properties of functionals and the notation used in this paper in connection with them. No attempt is made at mathematical rigor.

To say F is a functional of the function $q(\sigma)$ means that F is a number whose value depends on the *form* of the function $q(\sigma)$ (where σ is just a parameter used to specify the form of $q(\sigma)$). Thus,

$$F = \int_{-\infty}^{\infty} q(\sigma)^2 e^{-\sigma^2} d\sigma \qquad (1)$$

is a functional of $q(\sigma)$ since it associates with every choice of the function $q(\sigma)$ a number, namely the integral. Also, the area under a curve is a functional of the function representing the curve, since to each such function a number, the area is associated. The expected value of the energy in quantum mechanics is a functional of the wave function. Again,

$$F = q(0) \tag{2}$$

is a functional, which is especially simple because its value depends only on the value of the function $q(\sigma)$ at the one point $\sigma = 0$.

We shall write, if F is a functional of $q(\sigma)$, $F[q(\sigma)]$. A functional may have its argument more than one function, or functions of more than one parameter, as

$$F[x(t,s),\, y(t,s)] = \int_{-\infty}^{\infty}\int_{-\infty}^{\infty} x(t,s)y(t,s)\sin\omega(t-s)dtds\,.$$

A functional $F[q(\sigma)]$ may be looked upon as a function of an infinite number of variables, the variables being the value of the function $q(\sigma)$ at each point σ. If the interval of the range of σ is divided up into a large number of points σ_i, and the value of the function at these points is $q(\sigma_i) = q_i$, say, then approximately our functional may be written as a function of the variables q_i. Thus, in the case of equation (1) we could write, approximately,

$$F(\cdots q_i \cdots) = \sum_{i=-\infty}^{\infty} q_i^2 e^{-\sigma_i^2}(\sigma_{i+1} - \sigma_i)\,.$$

We may define a process analogous to differentiation for our functionals. Suppose the function $q(\sigma)$ is altered slightly to $q(\sigma) + \lambda(\sigma)$ by the addition of a small function $\lambda(\sigma)$. From our approximate viewpoint we can say that each of the variables is changed from q_i to $q_i + \lambda_i$. The function is thereby changed by an amount

$$\sum_i \frac{\partial F(\cdots q_i \cdots)}{\partial q_i}\lambda_i\,.$$

In the case of a continuous number of variables, the sum becomes an integral and we may write, to the first order in λ,

$$F[q(\sigma) + \lambda(\sigma)] - F[q(\sigma)] = \int K(t)\lambda(t)dt\,, \tag{3}$$

where $K(t)$ depends on F, and is what we shall call the functional derivative of F with respect to q at t, and shall symbolize, with Eddington,[5] by $\frac{\delta F[q(\sigma)]}{\delta q(t)}$. It is not simply $\frac{\partial F(\cdots q_i \cdots)}{\partial q_i}$ as this is in general infinitesimal, but is rather the sum of these $\frac{\partial F}{\partial q_i}$ over a short range of i, say from $i+k$ to $i-k$, divided by the interval of the parameter, $\sigma_{i+k} - \sigma_{i-k}$.

Thus we write,

$$F[q(\sigma)+\lambda(\sigma)] = F[q(\sigma)] + \int \frac{\delta F[q(\sigma)]}{\delta q(t)} \lambda(t) dt + \text{higher order terms in } \lambda\,. \tag{4}$$

For example, in equation (1) if we substitute $q+\lambda$ for q, we obtain

$$\begin{aligned} F[q+\lambda] &= \int [q(\sigma)^2 + 2q(\sigma)\lambda(\sigma) + \lambda(\sigma)^2] e^{-\sigma^2} d\sigma \\ &= \int q(\sigma)^2 e^{-\sigma^2} d\sigma + 2\int q(\sigma)\lambda(\sigma) e^{-\sigma^2} d\sigma \\ &\quad + \text{higher terms in } \lambda\,. \end{aligned}$$

Therefore, in this case, we have $\frac{\delta F[q]}{\delta q(t)} = 2q(t)e^{-t^2}$. In a similar way, if $F[q(\sigma)] = q(0)$, then $\frac{\delta F}{\delta q(t)} = \delta(t)$, where $\delta(t)$ is Dirac's delta symbol, defined by $\int \delta(t) f(t) dt = f(0)$ for any continuous function f.

The function $q(\sigma)$ for which $\frac{\delta F}{\delta q(t)}$ is zero for all t is that function for which F is an extremum. For example, in classical mechanics the action,

$$\mathscr{A} = \int L(\dot{q}(\sigma), q(\sigma)) d\sigma \tag{5}$$

is a functional of $q(\sigma)$. Its functional derivative is,

$$\frac{\delta \mathscr{A}}{\delta q(t)} = -\frac{d}{dt}\left\{\frac{\partial L(\dot{q}(t), q(t))}{\partial \dot{q}}\right\} + \frac{\partial L(\dot{q}(t), q(t))}{\partial q}\,. \tag{6}$$

If $\mathscr{A}$ is an extremum the right hand side is zero.

[5] A. S. Eddington, "The Mathematical Theory of Relativity" (1923) p. 139.
Editor's note: We have changed Eddington's symbol for the functional derivative to that now commonly in use.

2. The Principle of Least Action

For most mechanical systems it is possible to find a functional, $\mathscr{A}$, called the action, which assigns a number to each possible mechanical path, $q_1(\sigma), q_2(\sigma) \dots q_N(\sigma)$, (we suppose N degrees of freedom, each with a coordinate $q_n(\sigma)$, a function of a parameter (time) σ) in such a manner that this number is an extremum for an actual path $\bar{q}(\sigma)$ which could arise in accordance with the laws of motion. Since this extremum often is a minimum this is called the principle of least action. It is often convenient to use the principle itself, rather than the Newtonian equations of motion as the fundamental mechanical law. The form of the functional $\mathscr{A}[q_1(\sigma) \dots q_N(\sigma)]$ depends on the mechanical problem in question.

According to the principle of least action, then, if $\mathscr{A}[q_1(\sigma) \dots q_N(\sigma)]$ is the action functional, the equations of motion are N in number and are given by,

$$\frac{\delta \mathscr{A}}{\delta q_1(t)} = 0, \frac{\delta \mathscr{A}}{\delta q_2(t)} = 0, \dots, \frac{\delta \mathscr{A}}{\delta q_N(t)} = 0 \tag{7}$$

(We shall often simply write $\frac{\delta \mathscr{A}}{\delta q(t)} = 0$, as if there were only one variable). That is to say if all the derivatives of $\mathscr{A}$, with respect to $q_n(t)$, computed for the functions $\bar{q}_m(\sigma)$ are zero for all t and all n, then $\bar{q}_m(\sigma)$ describes a possible mechanical motion for the systems.

We have given an example, in equation (5), for the usual one dimensional problem when the action is the time integral of a Lagrangian (a function of position and velocity, only). As another example consider an action function arising in connection with the theory of action at a distance:

$$\mathscr{A} = \int_{-\infty}^{\infty} \left\{ \frac{m(\dot{x}(t))^2}{2} - V(x(t)) + k^2 \dot{x}(t)\dot{x}(t + T_0) \right\} dt \,. \tag{8}$$

It is approximately the action for a particle in a potential $V(x)$, and interacting with itself in a distant mirror by means of retarded and advanced waves. The time it takes for light to reach the mirror from the particle is assumed constant, and equal to $T_0/2$. The quantity

k^2 depends on the charge on the particle and its distance from the mirror. If we vary $x(t)$ by a small amount, $\lambda(t)$, the consequent variation in $\mathscr{A}$ is,

$$\delta\mathscr{A} = \int_{-\infty}^{\infty} \{m\dot{x}(t)\dot{\lambda}(t) - V'(x(t))\lambda(t) + k^2\dot{\lambda}(t)\dot{x}(t+T_0)$$

$$+ k^2\dot{\lambda}(t+T_0)\dot{x}(t)\}dt$$

$$= \int_{-\infty}^{\infty} \{-m\ddot{x}(t) - V'(x(t)) - k^2\ddot{x}(t+T_0)$$

$$- k^2\ddot{x}(t-T_0)\}\lambda(t)dt\,, \quad \text{by integrating by part}$$

so that, according to our definition (4), we may write,

$$\frac{\delta\mathscr{A}}{\delta x(t)} = -m\ddot{x}(t) - V'(x(t)) - k^2\ddot{x}(t+T_0) - k^2\ddot{x}(t-T_0)\,. \tag{9}$$

The equation of motion of this system is obtained, according to (7) by setting $\frac{\delta\mathscr{A}}{\delta x(t)}$ equal to zero. It will be seen that the force acting at time t depends on the motion of the particle at other time than t. The equations of motion cannot be described directly in Hamiltonian form.

3. Conservation of Energy. Constants of the Motion[6]

The problem we shall study in this section is that of determining to what extent the concepts of conservation of energy, momentum, etc., may be carried over to mechanical problems with a general form of action function. The usual principle of conservation of energy asserts that there is a function of positions at the time t, say, and of velocities of the particles whose value, for the actual motion of the particles, does not change with time. In our more general case however, the forces do not involve the positions of the particles only at one particular time, but usually a calculation of the forces requires

[6] This section is not essential to an understanding of the remainder of the paper.

a knowledge of the paths of the particles over some considerable range of time (see for example, Eq. (9)). It is not possible in this case generally to find a constant of the motion which only involves the positions and velocities at one time.

For example, in the theory of action at a distance, the kinetic energy of the particles is not conserved. To find a conserved quantity one must add a term corresponding to the "energy in the field". The field, however, is a functional of the motion of the particles, so that it is possible to express this "field energy" in terms of the motion of the particles. For our simple example (8), account of the equations of motion (9), the quantity,

$$E(t) = \frac{m(\dot{x}(t))^2}{2} + V(x(t)) - k^2 \int_t^{t+T_0} \ddot{x}(\sigma - T_0)\dot{x}(\sigma)d\sigma + k^2\dot{x}(t)\dot{x}(t+T_0)\,, \tag{10}$$

has, indeed, a zero derivative with respect to time. The first two terms represent the ordinary energy of the particles. The additional terms, representing the energy of interaction with the mirror (or rather, with itself) require a knowledge of the motion of the particle from the time $t - T_0$ to $t + T_0$.

Can we really talk about conservation, when the quantity conserved depends on the path of the particles over considerable ranges of time? If the force acting on a particle be $F(t)$ say, so that the particle satisfies the equation of motion $m\ddot{x}(t) = F(t)$, then it is perfectly clear that the integral,

$$I(t) = \int_{-\infty}^{t} [m\ddot{x}(t) - F(t)]\dot{x}(t)dt \tag{11}$$

has zero derivative with respect to t, when the path of the particle satisfies the equation of motion. Many such quantities having the same properties could easily be devised. We should not be inclined to say (11) actually represents a quantity of interest, in spite of its constancy.

The conservation of a physical quantity is of considerable interest because in solving problems it permits us to forget a great number of details. The conservation of energy can be derived from the laws of motion, but its value lies in the fact that by the use of it certain broad aspects of a problem may be discussed, without going into the great detail that is often required by a direct use of the laws of motion.

To compute the quantity $I(t)$, of equation (11), for two different times, t_1 and t_2 that are far apart, in order to compare $I(t_1)$ with $I(t_2)$, it is necessary to have detailed information of the path during the entire interval t_1 to t_2. The value of I is equally sensitive to the character of the path for all times between t_1 and t_2, even if these times lie very far apart. It is for this reason that the quantity $I(t)$ is of little interest. If, however, F were to depend on $x(t)$ only, so that it might be derived from a potential, (e.g.; $F = -V'(x)$), then the integrand is a perfect differential, and may be integrated to become $\frac{1}{2}m(\dot{x}(t))^2 + V(x(t))$. A comparison of I for two times, t_1 and t_2, now depends only on the motion in the neighborhood of these times, all of the intermediate details being, so to speak, integrated out.

We therefore require two things if a quantity $I(t)$ is to attract our attention as being dynamically important. The first is that it be conserved, $I(t_1) = I(t_2)$. The second is that $I(t)$ should depend only locally on the path. That is to say, if one changes the path at some time t' in a certain (arbitrary) way, the change which is made in $I(t)$ should decrease to zero as t' gets further and further from t. That is to say, we should like the condition $\frac{\delta I(t)}{\delta q_n(t')} \to 0$ as $|t - t'| \to \infty$ satisfied.[7]

[7] A more complete mathematical analysis than we include here is required to state rigorously just how fast it must approach zero as $|t - t'|$ approaches infinity. The proofs states herein are certainly valid if the quantities in (12) and (20) are assumed to become and remain equal to zero for values of $|t - t'|$ greater than some finite one, no matter how large it may be.

The energy expression (10) satisfies this criterion, as we have already pointed out. Under what circumstances can we derive an analogous constant of the motion for a general action function?

We shall, in the first place, impose a condition on the equations of motion which seems to be necessary in order that an integral of the motion of the required type exist. In the equation $\frac{\delta \mathscr{A}}{\delta q(t)} = 0$, which holds for an arbitrary time, t, we shall suppose that the influence of changing the path at time t' becomes less and less as $|t-t'|$ approaches infinity. That is to say, we require,

$$\frac{\delta^2 \mathscr{A}}{\delta q(t)\delta q(t')} \to 0 \quad \text{as} \quad |t - T'| \to \infty\,. \tag{12}$$

We next suppose that there exists a transformation (or rather, a continuous group of transformation) of coordinates, which we symbolize by $q_n \to q_n + x_n(a)$ and which leaves the action invariant (for example, the transformation may be a rotation). The transformation is to contain a parameter, a, and is to be a continuous function of a. For a equal to zero, the transformation should reduce to the identity, so that $x_n(0) = 0$. For very small a we may expand; $x_n(a) = 0 + ay_n + \ldots$. That is to say, for infinitesimal a, if the coordinates q_n are changed to $q_n + ay_n$ the action is left unchanged;

$$\mathscr{A}[q_n(\sigma)] = \mathscr{A}[q_n(\sigma) + ay_n(\sigma)]\,. \tag{13}$$

For example, if the form of the action is unchanged if the particles take the same path at a later time, we may take, $q_n(t) \to q_n(t+a)$. In this case, for small a, $q_n(t) \to q_n(t) + a\dot{q}_n(t) + \ldots$ so that $y_n = \dot{q}_n(t)$.

For each such continuous set of transformations there will be a constant of the motion. If the action is invariant with respect to change from $q(t)$ to $q(t + a)$, then an energy will exist. If the action is invariant with respect to the translation of all the coordinates (rectangular coordinates, that is) by the same distance, a, then a momentum in the direction of the translation may be derived. For rotations around an axis through the angle, the corresponding constant of the motion is the angular momentum around that axis. We may show this connection between the groups of transformations and the constants of the motion, in the following way: For small a, from (13), we shall have,

$$\mathscr{A}[q_n(\sigma)] = \mathscr{A}[q_n(\sigma) + ay_n(\sigma)]\,.$$

Expanding the left side with respect to the change in the coordinate $ay_n(\sigma)$, according to (4) to the first order in a we have,

$$\mathscr{A}[q_n(\sigma) + ay_n(\sigma)] = \mathscr{A}[q_n(\sigma)] + a\sum_{n=1}^{N}\int_{-\infty}^{\infty} y_n(t)\frac{\delta\mathscr{A}}{\delta q_n(t)}dt \tag{14}$$

and therefore, on account of (13),

$$\sum_{n=1}^{N}\int_{-\infty}^{\infty} y_n(\sigma)\frac{\delta\mathscr{A}}{\delta q_n(\sigma)}d\sigma = 0\,. \tag{15}$$

Now consider the quantity,

$$I(T) = \sum_{n=1}^{N}\int_{-\infty}^{T} y_n(\sigma)\frac{\delta\mathscr{A}}{\delta q_n(\sigma)}d\sigma\,. \tag{16}$$

On account of (15) we can also write,

$$I(T) = -\sum_{n=1}^{N}\int_{T}^{\infty} y_n(\sigma)\frac{\delta\mathscr{A}}{\delta q_n(\sigma)}d\sigma\,. \tag{17}$$

Consider the derivative of I with respect to T; $\frac{dI(T)}{dT} = \sum_n y_n(T)\frac{\delta\mathscr{A}}{\delta q_n(T)}$ According to the equations of motion (7), this is seen to vanish, so that $I(T)$ is independent of T for the real motion, and is therefore conserved. We must now prove, in order that it be acceptable as an important constant of the motion, that

$$\frac{\delta I(T)}{\delta q_m(t)} \to 0 \text{ as } |T-t| \to \infty \quad \text{for any } m\,. \tag{18}$$

Suppose first that $t > T$. Let us compute $\frac{\delta I(T)}{\delta q_m(t)}$ directly from equation (16), obtaining,

$$\frac{\delta I(T)}{\delta q_m(t)} = +\int_{-\infty}^{T} \sum_m \frac{\delta y_n(\sigma)}{\delta q_m(t)} \frac{\delta \mathscr{A}}{\delta q_m(\sigma)} d\sigma + \int_{-\infty}^{T} \sum_m y_n(\sigma) \frac{\delta^2 \mathscr{A}}{\delta q_m(t) \delta q_m(\sigma)} d\sigma \,. \tag{19}$$

Now we shall suppose that $y_n(\sigma)$ does not depend very much on values of $q_m(t)$ for times, t, far away from σ. That is to say we shall assume,[8]

$$\frac{\delta y_n(\sigma)}{\delta q_m(t)} \to 0 \quad \text{as} \quad |\sigma - t| \to \infty \,. \tag{20}$$

In the first integral then, since $t > T$, and since only values of σ less than T appear in the integrand, for all such values, $t - \sigma > t - T$. As $t - T$ approaches infinity, therefore, only terms in the first integral of (19) for which $t - \sigma$ approaches infinity appear. We shall suppose that $\frac{\delta y_n(\sigma)}{\delta q_m(t)}$ decreases sufficiently rapidly with increase in $t - \sigma$ that the integral of it goes to zero as $t - T$ becomes infinite. A similar analysis applies to the second integral of (19). Here the quantity $\frac{\delta^2 \mathscr{A}}{\delta q_m(t) \delta q_m(\sigma)}$ approaches zero because of our assumption (12), and we shall suppose this approach sufficiently rapid that the integral vanish in the limit.

Thus we have shown that $\frac{\delta I(T)}{\delta q_m(t)} \to 0$ as $t - T \to \infty$. To prove the corresponding relation for $T - t \to \infty$ one may calculate $\frac{\delta I(T)}{\delta q_m(t)}$ with $t < T$ from (17), and proceed in exactly the same manner. In this way we can establish the required relation (18). This then shows that $I(T)$ is an important quantity which is conserved.

A particularly important example is, of course, the energy expression. This is got by the transformation of displacing the time, as has

[8] In fact, for all practical cases which come to mind (energy momentum, angular momentum, corresponding to time displacement, translation, and rotation), $\frac{\delta y_n(\sigma)}{\delta q_m(t)}$ is actually zero if $\sigma \neq t$.

already been mentioned, for which $y_n(\sigma) = \dot{q}_n(\sigma)$. The energy integral may therefore be expressed, according to (16) (we have changed the sign), as,

$$E(T) = -\int_{-\infty}^{T} \sum_{n=1}^{N} \dot{q}_n(\sigma) \frac{\delta \mathscr{A}}{\delta q_m(\sigma)} d\sigma \,. \tag{21}$$

In our example (8) we would get from this formula,

$$\begin{aligned} E(T) = -\int_{-\infty}^{T} & \dot{x}(\sigma)[-m\ddot{x}(\sigma) - V'(x(\sigma)) - k^2\ddot{x}(\sigma + T_0) \\ & - k^2\ddot{x}(\sigma - T_0)]d\sigma \,, \end{aligned} \tag{22}$$

from which (10) has been derived by direct integration.

4. Particles Interacting through an Intermediate Oscillator

The problem we are going to discuss in this section, since it will give us a good example of a system for which only a principle of least action exists, is the following: Let us suppose we have two particles A and B which do not interact directly with each other, but there is a harmonic oscillator, O with which both of the particles A and B interact. The harmonic oscillator, therefore serves as an intermediary by means of which particle A is influenced by the motion of particle B and vice versa. In what way is this interaction through the intermediate oscillator equivalent to a direct interaction between the particles A and B, and can the motion of these particles, A, B, be expressed by means of a principle of least action, not involving the oscillator? (In the theory of electrodynamics this is the problem as to whether the interaction of particles through the intermediary of the field oscillators can also be expressed as a direct interaction at a distance.)

To make the problem precise, we let $y(t)$ and $z(t)$ represent coordinates of the particles A and B at the time t. Let the Lagrangians of the particles alone be designated by L_y and L_z. Let them each interact with the oscillator (with coordinate $x(t)$, Lagrangian

$\frac{m}{2}(\dot{x}^2 - \omega^2 x^2)$) by means of a term in the Lagrangian for the entire system, which is of the form $(I_y + I_z)x$, where I_y is a function involving the coordinates of atom A only, and I_z is some function of the coordinates of B. (We have assumed the interaction linear in the coordinate of the oscillator.)

We then ask: If the action integral for y, z, x, is

$$\int \left[L_y + L_z + \left(\frac{m\dot{x}^2}{2} - \frac{m\omega^2 x^2}{2} \right) + (I_y + I_z)x \right] dt \,, \tag{23}$$

is it possible to find an action $\mathscr{A}$, a functional of $y(t)$, $z(t)$, only, such that, as far as the motion of the particles A, B, are concerned, (i.e., for variations of $y(t)$, $z(t)$) the action $\mathscr{A}$ is a minimum?

In the first place, since the actual motion of the particles A, B, depends not only on y, z, initially (or at any other time) but also on the initial conditions satisfied by the oscillator, it is clear that $\mathscr{A}$ is not determined absolutely, but the form that $\mathscr{A}$ takes must have some dependence on the state of the oscillator.

In the second place, since we are interested in an action principle for the particles, we must consider variations of the motion of these particles from the true motion. That is, we must consider dynamically impossible paths for these particles. We thus meet a new problem; when varying the motion of the particle A and B, what do we do about the oscillator? We cannot keep the entire motion of the oscillator fixed, for that would require having this entire motion directly expressed in the action integral and we should be back where we started, with the action (23).

The answer to this question lies in the observation made above that the action must involve somehow some of the properties of the oscillator. In fact, since the oscillator has one degree of freedom it will require two numbers (e.g. position and velocity) to specify the state of the oscillator sufficiently accurately that the motion of the particles A and B is uniquely determined. Therefore in the action function for these particles, two parameters enter, which are arbitrary, and represent some properties of the motion of the oscillator. When the variation in the motion of the particles is taken, these quantities must be considered as constants, and thus, it is the properties of the oscillator described by these quantities which are considered fixed for the "impossible" motion of the particles. We should not be surprised to find that the action function for the particles depends on which properties of the oscillator are considered to be held fixed in the variation of y, and z. It is, on the other hand, somewhat unexpected that, as we shall see, not all possible conditions on the oscillator give rise to motions of y and z which are simply expressible in terms of an action principle. Let us see how this works out in detail.

We shall assume that the functions I_y and I_z are zero for times t greater than T and less than 0. We shall also assume (for simplicity only) that I_y is a function of t and $y(t)$ only, and does not depend on $\dot{y}(t)$. Similarly, I_z is to be independent of $\dot{z}(t)$. Then, from (23), the equation of motion of particle y is,

$$\frac{d}{dt}\left(\frac{\partial L_y}{\partial \dot{y}}\right) - \frac{\partial L_y}{\partial y} = \frac{\partial I_y}{\partial y} x(t)\,, \tag{24}$$

with a similar equation for z. That for the oscillator is,

$$m\ddot{x} + m\omega^2 x = [I_y(t) + I_z(t)]\,. \tag{25}$$

The solution of this last equation is, where we have written $\gamma = I_y + I_z$

$$x(t) = x(0)\cos\omega t + \dot{x}(0)\frac{\sin\omega t}{\omega} + \frac{1}{m\omega}\int_0^t \gamma(s)\sin\omega(t-s)ds\,. \tag{26}$$

This may be expressed in other ways, for example,

$$\begin{aligned} x(t) = {} & \frac{\sin\omega(T-t)}{\sin\omega T}\left[x(0) - \frac{1}{m\omega}\int_0^t \sin\omega s\gamma(s)ds\right] \\ & + \frac{\sin\omega t}{\sin\omega T}\left[x(T) - \frac{1}{m\omega}\int_t^T \sin\omega(T-s)\gamma(s)ds\right], \end{aligned} \tag{27}$$

or again,

$$x(t) = \frac{1}{\sin\omega T}[R_T \sin\omega t + R_0 \sin\omega(T-t)] + \frac{1}{2m\omega}\int_0^t \sin\omega(t-s)\gamma(s)ds - \frac{1}{2m\omega}\int_t^T \sin\omega(t-s)\gamma(s)ds\,, \tag{28}$$

where we have written

$$R_0 = \frac{1}{2}\left[x(0) + x(T)\cos\omega T - \dot{x}(T)\frac{\sin\omega T}{\omega}\right] \tag{29}$$

and

$$R_T = \frac{1}{2}\left[x(T) + x(0)\cos\omega T - \dot{x}(0)\frac{\sin\omega T}{\omega}\right]\,. \tag{30}$$

It is seen that R_T is the mean of the coordinate of the oscillator at time T and what that coordinate would have been at this time if the oscillator had been free and started with its actual initial conditions. Similarly, R_0 is the mean of the initial coordinate and what that coordinate would have had to be, were the oscillator free, to produce the actual final conditions at time T. Outside the time range 0 to T the oscillator is, of course, simply a free oscillator.

These expressions for $x(t)$ may be substituted into (24) and the corresponding equation for z, to obtain various expressions for the motion of the particles y, z; each expressed in terms of these particles and two parameters connected with the oscillator. For the expression (26) these parameters are $x(0)$ and $\dot{x}(0)$; for (27) they are $x(0)$ and $x(T)$; for (28) they are R_0 and R_T. We should like to determine whether these expressions could be obtained from an action for y and z only.

If the action be $\mathscr{A}$, then the expression (24) must be of the form $\frac{\delta\mathscr{A}}{\delta y(t)} = 0$. That is to say, we must have,

$$\frac{\delta\mathscr{A}}{\delta y(t)} = -\frac{d}{dt}\left(\frac{\partial L}{\partial \dot{y}}\right) + \left.\frac{\partial L}{\partial y}\right|_t + \left.\frac{\partial I_y}{\partial y}\right|_t \cdot x(t)\,. \tag{31}$$

We seek the solution of this expression for each expression we may write for $x(t)$.

Now an equation such as (31) (which is really an infinite set of equations, one for each value of t) does not always have a solution. One of the necessary requirements is, since $\frac{\delta}{\delta y(s)}\left(\frac{\delta\mathscr{A}}{\delta y(t)}\right) = \frac{\delta}{\delta y(t)}\left(\frac{\delta\mathscr{A}}{\delta y(s)}\right)$, that,

$$\frac{\delta}{\delta y(s)}\left[-\frac{d}{dt}\left(\frac{\partial L_y}{\partial \dot{y}}\right) + \left.\frac{\partial L_y}{\partial y}\right|_t + \left.\frac{\partial I_y}{\partial y}\right|_t \cdot x(t)\right]$$
$$= \frac{\delta}{\delta y(t)}\left[-\frac{d}{ds}\left(\frac{\partial L_y}{\partial \dot{y}}\right) + \left.\frac{\partial L_y}{\partial y}\right|_s + \left.\frac{\partial I_y}{\partial y}\right|_s \cdot x(s)\right].$$

This requires, therefore, if $s \neq t$, that $x(t)$ satisfy,

$$\left.\frac{\partial I_y}{\partial y}\right|_t \frac{\delta x(t)}{\delta y(s)} = \left.\frac{\partial I_y}{\partial y}\right|_s \frac{\delta x(s)}{\delta y(t)}\,. \tag{32}$$

For the expression (26), we have,

$$\begin{aligned} \frac{\delta x(t)}{\delta y(s)} &= \frac{1}{m\omega}\sin\omega(t-s)\cdot\left.\frac{\partial I_y}{\partial y}\right|_s \quad \text{if} \quad s < t \\ &= 0 \quad \text{if} \quad s > t\,. \end{aligned} \tag{33}$$

With this expression equation (32) is not satisfied so that we may conclude that no simple action function will describe the motion of the particles A and B if the initial position and velocity of the oscillator are considered as fixed.

On the other hand, since for the expression (27) we get,

$$\begin{aligned}\frac{\delta x(t)}{\delta y(s)} &= -\frac{\sin\omega(T-t)\sin\omega s}{m\omega\sin\omega T}\frac{\partial I_y}{\partial y}\bigg|_s \quad \text{if} \quad s<t \\ &= -\frac{\sin\omega(T-s)\sin\omega t}{m\omega\sin\omega T}\frac{\partial I_y}{\partial x}\bigg|_s \quad \text{if} \quad s>t\,,\end{aligned} \tag{34}$$

we may conclude, since (34) satisfies (32), that an action does exist if the oscillator has a given initial and a given final position. In fact, we may solve equation (31) with $x(t)$ expressed as in (27) and obtain, as an expression for the action,

$$\begin{aligned}\mathscr{A} &= \int_0^T [L_y + L_z]dt + \int_0^T \left[\frac{\sin\omega(T-t)x(0) + \sin\omega t x(T)}{\sin\omega T}\right]\gamma(t)dt \\ &\quad - \frac{1}{m\omega\sin\omega T}\int_0^T dt\int_0^t ds\cdot\sin\omega(T-t)\sin\omega s\gamma(s)\gamma(t)\,.\end{aligned} \tag{35}$$

The motion of the particles is given by this action principle by requiring that it be a minimum for variations of $y(t)$ and $z(t)$, where the quantities $x(0)$ and $x(T)$ are considered as fixed constants (for example, they might be zero).

In the case that $x(t)$ is given by the expression (28) with R_0 and R_T as constants, we find,

$$\begin{aligned}\frac{\delta x(t)}{\delta y(s)} &= \frac{1}{2m\omega}\sin\omega(t-s)\frac{\partial I_y}{\partial y}\bigg|_s \quad \text{if} \quad s<t \\ &= -\frac{1}{2m\omega}\sin\omega(t-s)\frac{\partial I_y}{\partial y}\bigg|_s \quad \text{if} \quad s>t\,,\end{aligned} \tag{36}$$

so that the relation (32) is satisfied for this case also. The action function for this case is,

$$\begin{aligned}\mathscr{A} &= \int_0^T [L_y + L_z]dt + \frac{1}{\sin\omega T}\int_0^T [R_T\sin\omega t + R_0\sin\omega(T-t)]\gamma(t)dt \\ &\quad + \frac{1}{2m\omega}\int_0^T\int_0^t \sin\omega(t-s)\gamma(t)\gamma(s)dsdt\,.\end{aligned} \tag{37}$$

This action is of particular interest inasmuch as in the special case $R_T = R_0 = 0$ the integrand no longer depends on T, and the limits of integration of 0 to T may be replaced without error by $-\infty$ and ∞ so that we obtain an action of the special form,

$$\mathscr{A} = \int_{-\infty}^{\infty} [L_y + L_z]dt + \frac{1}{2m\omega} \int_{-\infty}^{\infty} \int_{-\infty}^{t} \sin \omega(t-s)]\gamma(t)\gamma(s)dsdt\,. \tag{38}$$

If, now, after having passed to the limit $T = \infty$, we assume that $\gamma(t)$ and L_y, L_z do not depend on t explicitly, the substitution $y(t) \to y(t+a)$ and $z(t) \to z(t+a)$ does not alter the action function, so that an energy expression exists for this action. (In electrodynamics it leads to the half advanced plus half retard interaction used in the action at a distance theory.)

Exactly the same formulas result if it is that $\gamma(t)$ depends on the function $y(t)$, as any general functional. The action of the particles need not be the integral of a Lagrangian in the original form (23) either. If there are more than one intermediate oscillator, and if the oscillators are independent (i.e. if no two of the oscillators interact directly) the expressions for the action, with the oscillators eliminated are similar to (35), (37) and (38), except that there is a sum of interaction terms, one for each oscillator. Thus, if the frequency of the jth oscillator is ω_j, its mass m_j, the interaction with the particles given by γ_j and there were N oscillators, (38), for example, would read,

$$\mathscr{A} = \int_{-\infty}^{\infty} [L_y + L_z]dt + \sum_{j=1}^{N} \frac{1}{2m_j\omega_j} \int_{-\infty}^{\infty} \int_{-\infty}^{t} \sin \omega_j(t-s) \times \gamma_j(t)\gamma_j(s)dsdt\,. \tag{39}$$

By compounding terms of the form (39) many different types of interaction may be obtained. For example, the interaction of equation (8) would result from (39) if we had only one particle x, (instead of y and z), and an infinite number of oscillators of unit mass ($\frac{4}{\pi}\omega \sin \omega T d\omega$

oscillators with frequency ω in the range ω to $\omega + d\omega$), each interacting with the particle through the function $\gamma_j(t) = \dot{x}(t)$ (the same for each oscillator).

If we look a little more closely at the type of interaction of (38) we see that since $\gamma(t) = I_y(t) + I_z(t)$ the interaction contains terms of the form $I_y(t)I_y(s)$ and $I_z(t)I_z(s)$ as well as $I_y(t)I_z(s)$ and $I_z(t)I_y(s)$. Although the latter represent interactions between the particles A and B, the former type of term represents an interaction of particle A with itself, and particle B with itself, so to speak. What type of intermediate oscillator system will lead to interactions between particles, and to no interaction of a particle with itself?

This question is easily answered, when it is observed that if, in expression (39) for example, we take two oscillators, $j = 1$, and $j = 2$, so that $\omega_1 = \omega_2 = \omega$; $m_1 = -m_2 = m$; $\gamma_1 = I_y + I_z$; and $\gamma_2 = I_y - I_z$, since, $\gamma_1(s)\gamma_1(t) - \gamma_2(s)\gamma_2(t) = 2(I_y(s)I_z(t) + I_z(s)I_y(t))$, the interaction (39) could be written as,

$$\mathscr{A} = \int_{-\infty}^{\infty} [L_y + L_z]dt + \frac{1}{m\omega}\int_{-\infty}^{\infty}\int_{-\infty}^{t} \sin\omega(t-s) \times [I_y(s)I_z(t) + I_z(s)I_y(t)]dsdt \tag{40}$$

representing interactions of the particles with each other and having no "self-action" terms. Exactly the same procedure leads to the same results in the cases (35) and (37).

The original action, involving the oscillators, which leads to a form of this kind is, from what we have said,

$$\int_{-\infty}^{\infty}\left[L_y + L_z + (I_z + I_y)x_1 + (I_z - I_y)x_2 + \frac{m}{z}(\dot{x}_1^2 - \omega^2 x_1^2) - \frac{m}{2}(\dot{x}_2^2 - \omega^2 x_2^2)\right]dt\,.$$

This may be written, by putting $\eta_z = x_1 + x_2$ and $\eta_y = x_1 - x_2$,

$$\int_{-\infty}^{\infty}\left[L_y + L_z + I_z\eta_z + I_y\eta_y + \frac{m}{2}(\dot{\eta}_y\dot{\eta}_z - \omega^2\eta_y\eta_z)\right]dt\,.$$

This may be immediately generalized to the case where there are a number of particles y_k. The action,

$$\int_{-\infty}^{\infty}\left[\sum_k (L_{y_k} + I_{y_k}\eta_{y_k}) + \sum_k \sum_{l\neq k} \frac{m}{2}(\dot{\eta}_{y_k}\dot{\eta}_{y_l} - \omega^2 \eta_{y_k}\eta_{y_l})\right] dt$$

will lead to interactions only between pairs of particles k, l, no terms arising corresponding to the action of a particle on itself.

These action expressions will be of importance in the next part of the paper when we discuss them quantum mechanically. Starting with a system with a Hamiltonian, we have, at least classically, found a corresponding non-Hamiltonian action principle, by leaving out one member of the system. We have, in this, a way of checking a description which is intended as a generalization of quantum mechanics. We may start with a Hamiltonian system, where the quantum mechanics is well known, and show that by suitable elimination of the intermediate oscillator we get a system whose classical analogue obeys an action principle of the type (35) or (37). We do this on page 62.

III. Least Action in Quantum Mechanics

Classical mechanics is the limiting form of quantum mechanics when Planck's constant, $\hbar$, is considered as being vanishingly small. The classical system analogous to a quantum mechanical system (when such an analogue exists) may be mathematically exhibited most directly by letting $\hbar$ approach zero in the quantum mechanical equations.

The inverse problem, that of determining a quantum mechanical description of a system whose classical mechanical behaviour is known, may not be so easily solved. Indeed, the solution cannot be expected to be unique; witness, for example, the Klein-Gordon equation and Dirac's equation for the relativistic behaviour of an electron, both of which have the same classical analogue and quite different quantum mechanical consequences.

There exist, however, very useful rules applicable when the classical equations may be put into Hamiltonian form, and conjugate coordinates and momenta may be defined. These rules, leading to Schrödinger's equations, or Heisenberg's matrix formulation are too well known to require description here.[9]

For classical systems, such as those described in the first part of this paper, which in general have no Hamiltonian form, and for which conjugate momenta and coordinates cannot be defined, no satisfactory method of quantization has been given. In fact, the usual formulations of quantum mechanics use the concepts of Hamiltonian, and of momentum in such a direct and fundamental way that it would seem almost impossible to do without them.

A formulation of quantum mechanics will be described here which does not require the ideas of a Hamiltonian or momentum operator for its expression. It has, as its central mathematical idea, the analogue of the action integral of classical mechanics. It is a solution to the problem of the quantization of the classical theory of least action described in the first part of the paper.

A generalization of quantum mechanics which is to apply to a wider range of classical systems must satisfy at least two conditions. First, in the limit as $\hbar$ approaches zero the quantum mechanical equations must pass over into classical equations of motion applicable to systems in this widened range. And, second, they must be equivalent to the present formulations of quantum mechanics applicable to classical systems with Hamiltonia. The form of quantum mechanics to be described below does indeed satisfy both these criteria for systems for which a principle of least action exists. As an additional argument we shall show that the action principles arising

[9] They are discussed very satisfactorily by Dirac in his book, "The Principles of Quantum Mechanics" (Oxford 1935) on page 88, and on page 119.

in classical theory from the elimination of an intermediate harmonic oscillator arise in an analogous way in quantum mechanics.

1. The Lagrangian in Quantum Mechanics

A description of the proposed formulation of quantum mechanics might best begin by recalling some remarks made by Dirac[10] concerning the analogue of the Lagrangian and the action in quantum mechanics. These remarks bear so directly on what is to follow and are so necessary for an understanding of it, that it is thought best to quote them in full, even though it results in a rather long quotation. Speaking of the transformation function $(q'_t|q'_T)$ connecting the representations referring to two different times t and T, he says, "From the analogy between classical and quantum contact transformation ... we see that $(q'_t|q'_T)$ corresponds in the classical theory to $e^{iS/\hbar}$ where S is Hamilton's principal function for the time interval T to t, equal to the time-integral of the Lagrangian L,

$$S = \int_T^t L dt\,. \tag{21}$$

Taking an infinitesimal time interval t to $t+\delta t$, we see that $(q'_{t+\delta t}|q'_t)$ corresponds to $e^{\frac{iL\delta t}{\hbar}}$. This result gives probably the most fundamental quantum analogue for the classical Lagrangian function. It is preferable for the sake of analogy to consider the classical Lagrangian as a function of the coordinates at time t and the coordinates at time $t+\delta t$, instead of a function of the coordinates and velocities at time t.

There is an important action principle in classical mechanics concerning Hamilton's principal function (*21*). It says that this function remains stationary for small variations of the trajectory of the system which do not alter the end points, i.e. for small variations of the q's at all intermediate times between T and t with q_T and q_t fixed. Let us see what it corresponds to in the quantum theory.

---

[10] Dirac, "The Principles of Quantum Mechanics", p. 124–126.
Editor's note: Very similar material is contained in the Third Edition (1947), p. 125–130.

Put

$$\exp\left\{\frac{i}{\hbar}\int_{t_a}^{t_b} Ldt\right\} = \exp\left\{\frac{i}{\hbar}S(t_b, t_a)\right\} = B(t_b, t_a) \qquad (22)$$

so that $B(t_b, t_a)$ corresponds to $(q'_{t_b}|q'_{t_a})$ in the quantum theory. Now suppose the time interval $T \to t$ to be divided up into a large number of small time intervals $T \to t_1, t_1 \to t_2, \ldots, t_{m-1} \to t_m, t_m \to t$, by the introduction of a sequence of intermediate times $t_1, t_2, \ldots, t_m$. Then

$$B(t,T) = B(t,t_m)B(t_m,t_{m-1})\ldots B(t_2,t_1)B(t_1,T)\,. \qquad (23)$$

The corresponding quantum equation, which follows from the composition law ..., is

$$(q'_t|q'_T) = \iint\cdots\int (q'_t|q'_m)dq'_m(q'_m|q'_{m-1})dq'_{m-1}\cdots$$
$$\times (q'_2|q'_1)dq'_1(q'_1|q'_T)\,, \qquad (24)$$

q'_k being written for q'_{t_k} for brevity. At first sight there does not seem to be any close correspondence between (*23*) and (*24*). We must, however, analyse the meaning of (*23*) rather more carefully. We must regard each factor B as a function of the q's at the two ends of the time interval to which it refers. This makes the right-hand side of (*23*) a function, not only of q_t and q_T, but also of all the intermediate q's. Equation (*23*) is valid only when we substitute for the intermediate q's in its right-hand side their values for the real trajectory, small variations in which leave $S(t,T)$ stationary and therefore also, from (*22*), leave $B(t,T)$ stationary. It is the process of substituting these values for the intermediate q's which corresponds to the integrations over all values for the intermediate q's in (*24*). The quantum analogue of Hamilton's action principle is thus absorbed in the composition law (*24*) and the classical requirement that the values of the intermediate q's shall make $S(t,T)$ stationary corresponds to the condition in quantum mechanics that all values of the intermediate q's are important in proportion to their contribution to the integral in (*24*).

Let us see how (*23*) can be a limiting case of (*24*) for $\hbar$ small. We must suppose the integrand in (*24*) to be of the form $e^{iF/\hbar}$, where F is a function of $q'_T, q'_1, q'_2 \ldots q'_m, q'_t$ which remains continuous as $\hbar$ tends to zero, so that the integrand is a rapidly oscillating function when $\hbar$ is small. The integral of such a rapidly oscillating function will be extremely small, except for the contribution arising from a region in the domain of integration where comparatively large variations in the q'_k produce only very small variations in F. Such a region must be the neighborhood of a point where F is stationary for small variations of the q'_k. Thus the integral in (*24*) is determined essentially by the value of the integrand at a point where the integrand is stationary for small variations of the intermediate q''s, and so (*24*) goes over into (*23*)."

We may now point out that not only does $(q'_{t+\delta t}|q'_t)$ correspond to $\exp \frac{1}{\hbar}\left[L\left(\frac{q'_{t+\delta t}-q'_t}{\delta t}, q'_{t+\delta t}\right)\delta t\right]$, where $L(\dot{q}, q)$ is the Lagrangian for the classical system considered as a function of velocity and coordinate, but that often it is actually equal to it, within a normalization constant, in the limit as δt approaches zero. That is, to terms of order δt, if $\sqrt{g(q)}\,dq$ is the volume element in q-space

$$\begin{aligned}\psi(q'_{t+\delta t}, t+\delta t) &= \int (q'_{t+\delta t}|q'_t)\psi(q'_t, t)\sqrt{g(q'_t)}\,dq'_t \\ &= \int e^{\frac{i\delta t}{\hbar}L\left(\frac{q'_{t+\delta t}-q'_t}{\delta t}, q'_{t+\delta t}\right)}\psi(q'_t, t)\frac{\sqrt{g}\,dq'_t}{A(\delta t)}\end{aligned}$$

since $q'_{t+\delta t}$ and q'_t are just different variables it might be advantageous to avoid the subscripts, and to write;[11] $(Q = q'_{t+\delta t}, q = q'_t)$

$$\int e^{\frac{i\delta t}{\hbar}L\left(\frac{Q-q}{\delta t}, Q\right)}\psi(q, t)\frac{\sqrt{g(q)}dq}{A(\delta t)} = \psi(Q, t+\delta t) \tag{41}$$

---

[11] One could just as well write $L\left(\frac{q'_{t+\delta t}-q'_t}{\delta t}, q'_t\right)$ for $L\left(\frac{q'_{t+\delta t}-q'_t}{\delta t}, q'_{t+\delta t}\right)$ the difference being of a higher order of smallness than we are concerned with.

as $\delta t \to 0$ to the first order in δt. This is an integral equation to determine the wave function $\psi(Q, t+\delta t)$ for the system at the time t in terms of its value at time t. It therefore serves the same function as Schrödinger's equation, and indeed is equivalent to that equation if the normalization constant, $A(\delta t)$, a function of δt is chosen suitably.

To see how this comes about, we take the simplest case of a particle of mass m moving in one dimension in a force field of potential $V(x)$. Thus the classical Lagrangian function is $L = \frac{1}{2}m\dot{x}^2 - V(x)$. In accordance with equation (41), then the wave function for this system must satisfy (where we have written ε for δt) for infinitesimal ε, the equation,

$$\psi(x, t+\varepsilon) = \int e^{\frac{i\varepsilon}{\hbar}\{\frac{m}{2}(\frac{x-y}{\varepsilon})^2 - V(x)\}} \psi(y, t) \frac{dy}{A} \,. \tag{42}$$

Let us substitute $y = \eta + x$ in the integral;

$$\psi(x, t+\varepsilon) = \int e^{\frac{i}{\hbar}\{\frac{m\eta^2}{2\varepsilon} - \varepsilon V(x)\}} \psi(x+\eta, t) \frac{d\eta}{A} \,. \tag{43}$$

Only values of η close to zero will contribute to the integral, because, for small ε, other values of η make the exponential oscillate so rapidly that there will arise little contribution to the integral. We are therefore led to expand $\psi(x+\eta, t)$ in a Taylor series around $\eta = 0$, obtaining, after rearranging the integral,

$$\psi(x, t+\varepsilon) = \frac{e^{-\frac{i\varepsilon}{\hbar}V(x)}}{A} \int e^{\frac{i}{\hbar}\frac{m}{2\varepsilon}\eta^2} \left[\psi(x, t) + \eta \frac{\partial \psi(x, t)}{\partial x} + \frac{\eta^2}{2} \frac{\partial^2 \psi(x, t)}{\partial x^2} + \dots \right] d\eta \,.$$

Now, $\int_{-\infty}^{\infty} e^{\frac{im}{\hbar \cdot 2\varepsilon}\eta^2} \cdot d\eta = \sqrt{\frac{2\pi\hbar\varepsilon i}{m}}$ (see Pierces integral tables 487), and by differentiating both sides with respect to m, one may show

$$\int_{-\infty}^{\infty} \eta^2 \cdot e^{\frac{im}{\hbar \cdot 2\varepsilon}\eta^2} d\eta = \sqrt{\frac{2\pi\hbar\varepsilon i}{m}} \frac{\hbar\varepsilon i}{m} \,.$$

The integral with η in the integrand is zero since it is the integral of an odd function. Therefore,

$$\psi(x,t+\varepsilon) = \frac{\sqrt{\frac{2\pi\hbar\varepsilon i}{m}}}{A} e^{-\frac{i\varepsilon}{\hbar}V(x)} \left\{ \psi(x,t) + \frac{\hbar\varepsilon i}{m}\frac{\partial^2\psi}{2x^2} + \text{terms in } \varepsilon^2 \text{ etc.} \right\}. \tag{44}$$

The left hand side of this, for very small ε approaches $\psi(x,t)$ so that for the equality to hold we must choose,

$$A(\varepsilon) = \sqrt{\frac{2\pi\hbar\varepsilon i}{m}}. \tag{45}$$

Expanding both sides of (44) in powers of ε up to the first, we find,

$$\psi(x,t) + \varepsilon\frac{\partial\psi(x,t)}{\partial t} = \psi(x,t) - \frac{i\varepsilon}{\hbar}V(x)\psi(x,t) + \frac{\hbar i\varepsilon}{2m}\frac{\partial^2\psi}{\partial x^2},$$

and therefore,

$$-\frac{\hbar}{i}\frac{\partial\psi}{\partial t} = -\frac{\hbar^2}{2m}\frac{\partial^2\psi}{\partial x^2} + V(x)\psi$$

which is just Schrödinger's equation for the system in question.

This confirms the remark that equation (41) is equivalent to Schrödinger's differential equation for the wave function ψ.[12] Thus given a classical system describable by a Lagrangian which is a function of velocities and coordinates only, a quantum mechanical description of an analogous system my be written down directly, without first working out a Hamiltonian.

[12] We have, of course only proved the equivalence in a very special case. It is apparent, however, that the proof may be readily extended to any Lagrangian which is a quadratic function of the velocities, in a addition to, perhaps, some terms linear in the velocity arising from magnetic fields. The equivalence will be shown in a more general manner later.

If the problem involves more than one particle, or particles with more than one degree of freedom, equation (41) remains formally the same, if q and Q are considered to represent entire sets of coordinates, and $\int \sqrt{g}\, dq$ represents the volume integral over the space of these coordinates. The form of Schrödinger's equation which will be arrived at will be definite and will not suffer from the type of ambiguity one finds if one tries to substitute $\frac{\hbar}{i}\frac{\partial}{\partial q}$ for p_q in the classical Hamiltonian.

It will be important to consider the consequences of (41) when applied to connect the wave function at one time to its value at some finite later time.

Suppose we know the wave function $\psi(q_0, t_0)$ at time t_0, and desire it at time T. Dividing the time interval up into a very large number of small time intervals t_0 to t_1, t_1 to $t_2, \ldots$, t_m to T, and applying relation (41) to each time interval we obtain, if we write q_i for the coordinate associated with time t_i,

$$\psi(q_{i+1}, t_{i+1}) \simeq \int e^{\frac{i}{\hbar} L\left(\frac{q_{i+1}-q_i}{t_{i+1}-t_i}, q_{i+1}\right)\cdot(t_{i+1}-t_i)} \cdot \psi(q_i, t_i) \frac{\sqrt{g(q_i)}dq_i}{A(t_{i+1}-t_i)}\,. \tag{46}$$

Thus, by induction we find the relation,

$$\psi(Q,T) \cong \iint \cdots \int \exp\left\{\frac{i}{\hbar}\sum_{i=0}^{m}\left[L\left(\frac{q_{i+1}-q_i}{t_{i+1}-t_i}, q_{i+1}\right)\cdot(t_{i+1}-t_i)\right]\right\}$$
$$\times\, \psi(q_0, t_0)\frac{\sqrt{g_0}dq_0\sqrt{g_1}dq_1, \ldots, \sqrt{g_n}dq_m}{A(t_1-t_0)\cdot A(t_2-t_1)\cdot \ldots \cdot A(T-t_m)}\,, \tag{47}$$

where in the sum we shall write Q for q_{m+1}, and T for t_{m+1}. In the limit as we take finer and finer subdivisions of the interval t_0 to T and thus make an ever increasing number of successive integrations, the expression on the right side of (47) becomes equal to $\psi(Q,T)$. The sum in the exponential resembles $\int_{t_0}^{T} L(\dot{q}, q)dt$ with the integral written as a Riemann sum.

In a similar manner we can compute $\psi(q_0, t_0)$ in terms of the wave function at the later time $T = t_{m+1}$, by the equation;

$$\psi^*(q_0, t_0) = \iint \cdots \int \psi^*(q_{m+1}, t_{m+1}) \times \exp\left\{\frac{i}{\hbar}\sum_{i=0}^{m}\left[L\left(\frac{q_{i+1}-q_i}{t_{i+1}-t_i}, q_{i+1}\right)\cdot(t_{i+1}-t_i)\right]\right\} \times \frac{\sqrt{g_{m+1}}dq_{m+1}\cdots\sqrt{g_1}dq_1}{A(t_{m+1}-t_m)\cdots A(t_1-t_0)}\,. \tag{48}$$

2. The Calculation of Matrix Elements in the Language of a Lagrangian

Suppose we wish to compute the average value of some function $f(q)$ of the coordinates, at the time t_0, which we shall call $\langle f(q_0)\rangle$;

$$\langle f(q_0)\rangle = \int \psi^*(q_0, t_0) f(q_0)\psi(q_0, t_0)\sqrt{g_0}dq_0\,. \tag{49}$$

Let us try to express this in terms of the wave function at some more future time, $t = t_{m+1}$, by equation (48), and in terms of the wave function $\psi(q_{-m'}, t_{-m'})$ at some earlier time $t = t_{-m'}$ by an equation analogous to (47) (we shall let negative indices stand for times earlier than t_0). We obtain

$$\langle f(q_0)\rangle = \iint \cdot \int \psi^*(q_{m+1}, t_{m+1}) \times \exp\left\{\frac{i}{\hbar}\sum_{i=-m'}^{m}\left[L\left(\frac{q_{i+1}-q_i}{t_{i+1}-t_i}, q_{i+1}\right)\cdot(t_{i+1}-t_i)\right]\right\}\cdot f(q_0) \cdot\psi(q_{-m'}, t_{-m'})\cdot\frac{\sqrt{g}\,dq_{m+1}\ldots\sqrt{g}\,dq_0\sqrt{g}\,dq_{-1}\ldots\sqrt{g}dq_{-m'}}{A(t_{m+1}-t_m)\ldots A(t_0-t_{-1})\ldots A(t_{-m'+1}-t_{-m'})}\,. \tag{50}$$

We shall be dealing with expressions of this form to a considerable extent, so that we shall make some remarks concerning it. In the first place, we will always take $t_{-m'}$ to be a fixed time T_1 very far in the past, and t_{m+1} to be T_2, a time very far in the future. Secondly, we may generalize somewhat by taking the wave function at T_2 arbitrarily, say χ, and not necessarily the same as what the wave function ψ chosen at time T_1 would become at this later time T_2. In this way we shall have an expression more like a matrix element than an average. We shall use the symbol $\langle\chi|f(q_0)|\psi\rangle$ to represent this quantity. We shall continue to call it simply the average of f even though it more closely resembles a matrix element. We may, of course calculate quantities such as $\langle\chi|f(q_1)|\psi\rangle$, which correspond to the average value of $f(q)$ at the slightly later time, t_1, merely by replacing the $f(q_0)$ appearing in the integral on the right hand side of (50) with $f(q_1)$. Thus, we may find the time rate of change of the average value of $f(q)$, as,

$$\frac{d}{dt}\langle\chi|f(q)|\psi\rangle = \frac{\langle\chi|f(q_1)|\psi\rangle - \langle\chi|f(q_0)|\psi\rangle}{t_1 - t_0} = \langle\chi|\frac{f(q_1) - f(q_0)}{t_1 - t_0}|\psi\rangle\,, \tag{51}$$

where the last expression means that we are to replace $f(q_0)$ in the integral of (50) by $\frac{f(q_1)-f(q_0)}{t_1-t_0}$. We may say then, that the symbol $\langle\chi|\quad|\psi\rangle$ means that the quantity inside is to be multiplied by an exponential of the form of equation (50), by a wave function χ at T_2, and one, ψ, at T_1, and integrated over all coordinates. Finally, the limit as the subdivisions of time become finer and finer is to be taken. We shall, for a while, disregard which wave functions are put into the expression, and shall simply write, $\langle|f|\rangle$, but we shall discuss this later. In this way we can define the average of $F(q(\sigma)]$ where F is any functional[13] at all of $q(\sigma)$. We need merely to express the functional approximately as a function of the values, q_i, of q at the points t_i, place this function in the integral of (50) and pass to the limit.

[13] Properties of functionals and our notation with regard to them is described in the first section (p. 7) of the paper.

3. The Equations of Motion in Lagrangian Form

Let us now consider some functional, which, expressed in terms of q_i, the values of q at the times t_i is $F(q_i)$, - that is to say, a function of $\dots q_{-1}, q_0, q_1, q_2 \dots$ etc. Let us calculate $\left\langle \left| \frac{1}{\sqrt{g_n}} \frac{\partial(\sqrt{g_n}F)}{\partial q_n} \right| \right\rangle$. Replacing f by $\frac{1}{\sqrt{g}} \frac{\partial \sqrt{g}F}{\partial q_k}$ in (50) we see that we may integrate by parts with respect to q_k. The integrated part may be assumed to vanish, for if we consider that the integrations over the other q's were performed the remaining integrand would be similar to the square of the wave function at time t_k, which presumably vanishes at infinity. We find after integrating by parts, an expression similar to (50) but having a different form for $f(q_0)$; namely we find,

$$\left\langle \left| \frac{1}{\sqrt{g(q_k)}} \frac{\partial(\sqrt{g(q_k)} \cdot F)}{\partial q_k} \right| \right\rangle$$
$$= -\frac{i}{\hbar} \left\langle \left| F \cdot \frac{\partial}{\partial q_k} \left\{ \sum_{i=-m'}^{m} \left[L\left(\frac{q_{i+1} - q_i}{t_{i+1} - t_i}, q_{i+1} \right) \cdot (t_{i+1} - t_i) \right] \right\} \right| \right\rangle . \tag{52}$$

Thus performing the indicated differentiation we obtain,

$$\left\langle \left| \frac{1}{\sqrt{g}} \frac{\partial(\sqrt{g}F)}{\partial q_k} \right| \right\rangle$$
$$= \frac{i}{\hbar} \left\langle \left| F \left\{ L_{\dot{q}} \left[\frac{q_{k+1} - q_k}{t_{k+1} - t_k}, q_{k+1} \right] - L_{\dot{q}} \left(\frac{q_k - q_{k-1}}{t_k - t_{k-1}}, q_k \right) \right. \right. \right.$$
$$\left. \left. \left. - (t_k - t_{k+1}) \cdot L_q \left(\frac{q_k - q_{k-1}}{t_k - t_{k-1}}, q_k \right) \right\} \right| \right\rangle , \tag{53}$$

where we have written $L_{\dot{q}}$ for the function $\frac{\partial L}{\partial \dot{q}}$, and L_q for $\frac{\partial L}{\partial q}$. The expression in $\{ \; \}$ may be remembered most easily by noting that its limit for infinitesimal subdivisions is

$$\left[\frac{d}{dt} \left(\frac{\partial L}{\partial \dot{q}} \right) - \frac{\partial L}{\partial q} \right] dt \, .$$

This relation (53) is fundamental in that, when compared to corresponding expressions in the usual form of quantum mechanics, it contains, as we shall see, in one equation, both the equations of motion and the commutation rules for p and q. The way this comes about can be seen most clearly by applying equation (53) to the simple example with $L = m\frac{\dot{x}^2}{2} - V(x)$. It then reads, ($\sqrt{g} = 1$)

$$\left\langle \left| \frac{\partial F}{\partial x_k} \right| \right\rangle = \frac{i}{\hbar} \left\langle \left| F \cdot \left\{ m \left(\frac{x_{k+1} - x_k}{t_{k+1} - t_k} \right) \right.\right.\right.$$
$$\left.\left.\left. - m \left(\frac{x_k - x_{k-1}}{t_k - t_{k-1}} \right) + (t_k - t_{k-1}) V'(x_k) \right\} \right| \right\rangle . \quad (54)$$

If $F = x_k$ then (54) becomes;

$$\langle |1| \rangle = \frac{i}{\hbar} \left\langle \left| m \left(\frac{x_{k+1} - x_k}{t_{k+1} - t_k} \right) x_k - x_k \cdot m \left(\frac{x_k - x_{k-1}}{t_k - t_{k-1}} \right) \right.\right.$$
$$\left.\left. + x_k (t_k - t_{k-1}) V'(x_k) \right| \right\rangle .$$

In the limit of fine subdivision of the time, since as $t_k - t_{k-1} \to 0$ the last term becomes unimportant, we can write, supposing as usual that in all these equations the limit is to be taken,

$$\left\langle \left| \left(m \frac{x_{k+1} - x_k}{t_{k+1} - t_k} \right) x_k - x_k \left(m \frac{x_k - x_{k-1}}{t_k - t_{k-1}} \right) \right| \right\rangle = \frac{\hbar}{i} \langle |1| \rangle . \quad (55)$$

This is equivalent to the statement, in the ordinary notation of quantum mechanics, that the average value of $pq - qp$ is equal to the average value of $\frac{\hbar}{i} I$. The order of the factors in the usual mechanics here shows up as the order of the terms in time. (Exact relations by which formulas of the notation of equation (55) can be translated into relations in the more usual notation will be given in a later section (page 40)).

Again by equation (53) for $F = x_{k+3}$, say, we find,

$$\left\langle \left| x_{k+3} \left(m \frac{x_{k+1} - x_k}{t_{k+1} - t_k} \right) - x_{k+3} \left(m \frac{x_k - x_{k-1}}{t_k - t_{k-1}} \right) \right| \right\rangle = 0 \text{ in the limit.} \quad (56)$$

Thus, the difference of two successive momentum measurements followed by a position measurement, multiplied and averaged, is infinitesimal since the two successive momentum measurements give the same result — but if the position measurement occurs *between* the momentum measurements, in time, the results is no longer small, the position measurement having disturbed the momentum between the times the momentum was measured. (Of course, these quantities cannot actually be looked upon as averages of quantities in the classical sense because of the i in the expressions.)

In Eq. (55) we may replace $\left\langle\middle|x_k \cdot m\left(\frac{x_k - x_{k-1}}{t_k - t_{k-1}}\right)\middle|\right\rangle$ by its value a moment later, namely, $\left\langle\middle|x_{k+1} \cdot m\left(\frac{x_{k+1} - x_k}{t_{k+1} - t_k}\right)\middle|\right\rangle$ without changing the value of the expression by a finite amount. Thus (55) may be rearranged to read,

$$\langle|(x_{k+1} - x_k)^2|\rangle = -\frac{\hbar}{mi}(t_{k+1} - t_k) \cdot \langle|1|\rangle\,. \tag{57}$$

This describes the well known fact that a wave packet spreads parabolically in time from a point, and that although the average value of the displacement of a particle in the time dt is vdt, where v is the mean velocity, the mean value of the square of this displacement is not of order dt^2, but only of order dt. We mention it here to point out that although the form for average velocity, is, from equation (51), $\left\langle\middle|\frac{x_{k+1} - x_k}{t_{k+1} - t_k}\middle|\right\rangle$, the average kinetic energy, for example, must be written $\left\langle\middle|\frac{m}{2}\left(\frac{x_{k+1} - x_k}{t_{k+1} - t_k} \cdot \frac{x_k - x_{k-1}}{t_k - t_{k-1}}\right)\middle|\right\rangle$ rather than $\left\langle\middle|\frac{m}{2}\left(\frac{x_{k+1} - x_k}{t_{k+1} - t_k}\right)^2\middle|\right\rangle$. The latter is infinite.

If, in (54), we had chosen for F the expression $G_1 x_k G_2$, where G_1 is any function of the coordinates, x_j, belonging to times t_j later than t_k ($t_j > t_k$), and G_2 is any function of the coordinates belonging to times earlier than t_k, we would have found, in place of equation (55), the relation,

$$\left\langle\middle|G_1\left[\left(m\frac{x_{k+1} - x_k}{t_{k+1} - t_k}\right) \cdot x_k - x_k \cdot \left(m\frac{x_k - x_{k-1}}{t_k - t_{k-1}}\right)\right]G_2\middle|\right\rangle$$
$$= \frac{\hbar}{i}\langle|G_1 G_2|\rangle\,. \tag{58}$$

This is equivalent to the usual relation among averages,

$$\langle|G_1(pq - qp)G_2|\rangle = \frac{\hbar}{i}\langle|G_1G_2|\rangle\,.$$

Since G_1 and G_2 are arbitrary functions of their coordinates, we may think of equation (58) as equivalent to the operator equation $pq - qp = \frac{\hbar}{i}$.

Replacing F simply by G_1G_2 with G_1, G_2 defined as before, equation (54) becomes after dividing through by $(t_k - t_{k-1})$ since the left side is zero,

$$\left\langle\left|G_1\left[\frac{m\left(\frac{x_{k+1}-x_k}{t_{k+1}-t_k}\right) - m\left(\frac{x_k - x_{k-1}}{t_k - t_{k-1}}\right)}{t_k - t_{k-1}} + V'(x_k)\right]G_2\right|\right\rangle = 0\,. \quad (59)$$

This is equivalent to the operator equation, in the usual notation of quantum mechanics, which is the quantum analogue of Newton's law of motion, namely, $m\ddot{x} + V'(x) = 0$. This law and the commutation rules are of course equivalent to the commutation rules and the rules $HF - FH = \frac{\hbar}{i}\dot{F}$ with $H = \frac{1}{2m}p^2 + V(x)$ in the usual formulation. Thus equations (58) and (59) state all that is needed to completely solve the problem for this system, and hence equation (54), or its generalization, equation (53), from which they can be derived is all that is required.

Even though we are going to extend this to problems for which no Hamiltonian exists, it may still be of interest to describe the Hamiltonian from our point of view, when one does exist. Let us consider the average of any functional $F(q_i)$. To calculate the rate of change with the time of this quantity we may use a relation analogous to (51). Another method is as follows; suppose the variables q_i which appear in F are limited to indices between the times t_l and $t_{l'}$ $(l > l')$. That is to say, F is a functional of only the variables q_{l-1} to $q_{l'+1}$. (In the limit as our subdivisions in time become infinite, l, l' may become infinite, but we want t_l and $t_{l'}$ to remain bounded, so that F covers only a finite span of time). Now, if in our expression (50) for the average value of F the values of the times t_i for i between

l and l' were increased by a constant small amount δ it would be equivalent to calculating our F at a time δ later, with, however, the same wave functions kept fixed at fixed times, T_2 and T_1. This will give us just $\langle|F|\rangle + \delta\frac{d}{dt}\langle|F|\rangle$ to the first order in δ. Therefore, $\frac{d}{dt}\langle|F|\rangle$ is the derivative with respect to δ of the quantity we get by so augmenting the time variables. To compute the indicated quantities we look at equation (50) and notice that if all the times were altered as indicated, the only change made in the formula would be to replace $t_l - t_{l-1}$ by $t_l - t_{l-1} - \delta$ and $t_{l'+1} - t_{l'}$ by $t_{l'+1} - t_{l'} + \delta$. Doing this, and taking the derivative with respect to δ, we find,

$$\begin{aligned}\frac{d}{dt}\langle|F|\rangle = \frac{i}{\hbar}\Big\langle\Big|\Big\{&L_{\dot{q}}\left(\frac{q_l - q_{l-1}}{t_l - t_{l-1}}, q_l\right)\cdot\frac{q_l - q_{l-1}}{t_l - t_{l-1}}\\ &- L\left(\frac{q_l - q_{l-1}}{t_l - t_{l-1}}, q_l\right) + \frac{\hbar}{i}\alpha(t_l - t_{l-1})\Big\}F\\ &- F\Big\{L_{\dot{q}}\left(\frac{q_{l'+1} - q_{l'}}{t_{l'+1} - t_{l'}}, q_{l'+1}\right)\cdot\frac{q_{l'+1} - q_{l'}}{t_{l'+1} - t_{l'}}\\ &- L\left(\frac{q_{l'+1} - q_{l'}}{t_{l'+1} - t_{l'}}, q_{l'+1}\right) + \frac{\hbar}{i}\alpha(t_{l'+1} - t_{l'})\Big\}\Big|\Big\rangle, \end{aligned} \tag{60}$$

where $\alpha(\delta t) = \frac{d}{d(\delta t)}\ln A(\delta t)$. Equation (60) only applies if F contains only coordinates between q_l and $q_{l'}$, and does not involve the time explicitly. If F does involve the time a term $\sum_{i=l'}^{l}\langle|\frac{\partial F}{\partial t_i}|\rangle$ should be added to the right side.

We may compare (60) to the usual relation $\frac{d}{dt}\langle F\rangle = \frac{i}{k}\langle HF - HF\rangle$ and see that the analogue of H is the expression in the $\{\ \}$ in this equation. The terms $\frac{\hbar}{i}\alpha(t_l - t_{l-1})$ arise from the differentiation of the normalizing factor, A. They serve to keep the expression for H finite in the limit of infinitesimal subdivision, in spite of the fact that it may contain terms of the form $\langle|\left(\frac{q_l - q_{l-1}}{t_l - t_{l-1}}\right)^2|\rangle$ whose magnitude we have already discussed. For example, in view of (57) for the simple Lagrangian $\frac{m\dot{x}^2}{2} - V(x)$, it is seen that α must be $\frac{1}{2(t_l - t_{l-1})}$ and that, therefore, $A(t_l - t_{l-1}) = \text{const.}\sqrt{t_l - t_{l-1}}$, as we have already found for this case (Eq. (45)).

4. Translation to the Ordinary Notation of Quantum Mechanics

What we have been doing so far is no more than to reexpress ordinary quantum mechanics in a somewhat different language. In the next few pages we shall require this altered language in order to describe the generalization we are to make to systems without a simple Lagrangian function of coordinates and velocities. Before we do this, it is perhaps worth while to show how the relations we have derived up to this point, for systems possessing a Hamiltonian, H say, may be most readily translated into the more usual notation.

The usual expression for the wave function at time t_2 in terms of the wave function at time t_1 is given by means of the relation,

$$\psi_{t_2} = e^{\frac{i}{\hbar}H(t_2-t_1)}\psi_{t_1} \,. \tag{61}$$

Thus the matrix element between a state which, at time T_2, possesses the wave function χ, and the state represented at time T_1 by the wave function ψ, of an operator F considered acting at the time t_0 is

$$\int \bar{\chi} e^{\frac{i}{\hbar}H(T_2-t_0)} F e^{\frac{i}{\hbar}(t_0-T_1)H} \psi d\,\text{Vol} \,. \tag{62}$$

Thus, for example, the average coordinate at time t_0 between these states, which is what we have been calling $\langle\chi|q_0|\psi\rangle$ is expressible as

$$\langle\chi|q_0|\psi\rangle = \int \bar{\chi} e^{\frac{i}{\hbar}(T_2-t_0)H} q e^{\frac{i}{\hbar}(t_0-T_1)H} \psi d\,\text{Vol} \,. \tag{63}$$

Similarly, $\langle\chi|q_1|\psi\rangle$ is the matrix element of the operator $e^{+\frac{i}{\hbar}(T_2-t_1)H} q e^{\frac{i}{\hbar}H(t_1-T_1)}$, so that as $t_1 - t_0 \to 0$ the quantity,

$$\langle\chi|\frac{q_1-q_0}{t_1-t_0}|\psi\rangle = \int \bar{\chi} e^{\frac{i}{\hbar}H(T_2-t_1)} \frac{\left\{q - e^{\frac{i}{\hbar}(t_1-t_0)H} q e^{-\frac{i}{\hbar}(t_1-t_0)H}\right\}}{t_1-t_0}$$
$$\times\, e^{\frac{i}{\hbar}(t_1-T_1)H}\psi d\,\text{Vol}$$

becomes $\int \bar{\chi} e^{+\frac{i}{\hbar}H(T_2-t_1)} \cdot \frac{i}{\hbar}(Hq - qH) e^{\frac{i}{\hbar}H(t_1-T_1)} \psi d\,\mathrm{Vol}$, and is therefore the average value of the operator $\frac{i}{\hbar}(Hq - qH)$, and represents velocity, as we have said. Other functional expressions can be made to correspond to operators. For example, if $t_m > t_l$,

$$\langle \chi | q_m q_l^2 | \psi \rangle = \int \bar{\chi} e^{\frac{i}{\hbar}H(T_2-t_m)} q e^{\frac{i}{\hbar}H(t_m-t_l)} q^2 e^{\frac{i}{\hbar}H(t_l-T_1)} \psi d\,\mathrm{Vol} \tag{64}$$

while if $t_m < t_l$ it is given by

$$\langle \chi | q_m q_l^2 | \psi \rangle = \int \bar{\chi} e^{\frac{i}{\hbar}H(T_2-t_l)} q^2 e^{\frac{i}{\hbar}H(t_l-t_m)} q e^{\frac{i}{\hbar}H(t_m-T_1)} \psi d\,\mathrm{Vol}\,. \tag{65}$$

For example, equation (55) when translated becomes, if $H = \frac{1}{2m}p^2 + V(x)$,

$$\int \bar{\chi} e^{\frac{i}{\hbar}H(T_2-t_k)} \cdot \frac{mi}{\hbar}\{(Hx - xH)x - x(Hx - xH)\} e^{\frac{i}{\hbar}H(t_k-T_1)} \psi d\,\mathrm{Vol}$$
$$= \frac{\hbar}{i} \int \bar{\chi} e^{\frac{i}{\hbar}H(T_2-T_1)} \psi d\,\mathrm{Vol}$$

which, of course, is correct. The left side of equation (56) when translated with the aid of equations (64) and (65) does indeed vanish in the limit as the division of time into intervals becomes finer and finer.

As a further example we may mention the equivalence between $\langle | \dot{x}(t) \cdot f(x(t)) | \rangle$ and the average of

$$e^{\frac{i}{\hbar}H(T_2-t)} \left[\frac{1}{2m}(pf(x) + f(x)p) \right] e^{\frac{i}{\hbar}H(t-T_1)} \tag{66}$$

as may be seen most easily by considering the average of $\frac{x_{i+1}-x_{i-1}}{t_{i+1}-t_{i-1}} f(x_i)$ according to obvious generalizations of equations (64) and (65).

5. The Generalization to any Action Function

We now make the generalization to the case when the classical action need not be of the form $\mathscr{A} = \int L(\dot{q}, q)dt$, but is some other more general functional of $q(\sigma)$. In equation (50), as has already been remarked, the phase of the exponential is just $\frac{i}{\hbar}\int L(\dot{q}, q)dt$ written as a Riemann sum due to our subdivision of the time into finite, but small, intervals. The obvious suggestion is, then, to replace this exponent by $\frac{i}{\hbar}$ times the more general action. The action must of course first be expressed in an approximate way in terms of q_i, t_i, in such a way that as the subdivision becomes finer and finer it more nearly approaches the action expressed as a functional of $q(t)$.

In order to get a clearer idea of what this will lead to, let us choose a simple action function to keep in mind, for which no Hamiltonian exists. We may take,

$$\mathscr{A} = \int_{-\infty}^{\infty} \left\{ \frac{m\dot{x}(t)^2}{2} - V(x(t)) + k^2\dot{x}(t)\dot{x}(t+T_0) \right\} dt \tag{67}$$

which is an approximate action function for a particle in a potential $V(x)$ and which also interacts with itself in a mirror, by means of half advanced and half retarded waves, exactly as in (8).

In the expression of equation (50) the integral of L extends only over a finite time range T_1 to T_2. Our action (67) is meaningless for a finite time range. In fact if we were to integrate over the range from T_1 to T_2 the action might still depend on values of $x(t)$ outside this range.

This difficulty may be circumvented by altering our mechanical problem. We may assume that at a certain very large positive time T_2, and at a large negative time T_1, all of the interactions (e.g., the charges) have gone to zero and the particles are just a set of free particles (or at least their motion is describable by a Lagrangian). We may then put wave functions, χ and ψ, for these times, when the particles are free, into (50). (We might then suppose that the motion in the actual problem may be a limit of the motion as these times T_1 and T_2 move out to infinity). We therefore compute, by analogy to (50) the quantity

$$\langle\chi|F|\psi\rangle = \int \chi^*(q_{T_2}) \exp\left\{\frac{i}{\hbar}\mathscr{A}(q_{T_2}\ldots q_2, q_1, q_0, q_{-1}, \ldots q_{T_1})\right\}$$
$$\cdot F(\ldots q_1, q_0 \ldots) \cdot \psi(q_{T_1}) \frac{\sqrt{g}dq_{T_2}\cdots\sqrt{g}dq_{T_1}}{A(T_2 - t_m)\ldots A(t_{-m'} - T_1)} \,. \quad (68)$$

In the limit as $\hbar \to 0$ this will lead to the classical action principle, $\delta\mathscr{A} = 0$, in the way outlined by Dirac (see page 27) since nothing is altered to invalidate that argument.

We may obtain the fundamental relation of our quantum mechanics, analogous to equation (52), by integrating the formula for the average of $\frac{1}{\sqrt{g}}\frac{\partial(\sqrt{g}F)}{\partial q_k}$ by parts, to obtain,

$$\langle\chi|\frac{1}{\sqrt{g(q_k)}}\frac{\partial(\sqrt{g(q_k)}F)}{\partial q_k}|\psi\rangle = -\frac{i}{\hbar}\langle\chi|F \cdot \frac{\partial\mathscr{A}}{\partial q_k}|\psi\rangle \,. \quad (69)$$

As we have remarked above, this contains the analogue of the equations of motion as well as the quantum commutation rules.

6. Conservation of Energy. Constants of the Motion

Because of the importance in ordinary quantum mechanics of operators which correspond to classical constants of motion, we shall mention briefly the analogue of these operators in our generalized formulation. Since these are not needed for the remainder of the paper, they have not been studied in detail.

The notation will be as in the classical case described in section 3, of the first part of the paper. The general discussion given there applies equally well in this case, so that we shall not repeat it. We will suppose, for simplicity that there is only one coordinate, $q(\sigma)$, instead of the N coordinates $q_n(\sigma)$.

From the equations of motion (69) we can verify directly that,

$$\langle\chi|\mathscr{F}\cdot\sum_{t_i=\bar{t}_1}^{t_i=\bar{t}_2}\left[y_i\frac{\partial\mathscr{A}}{\partial q_i}+\frac{\hbar}{i}\frac{1}{\sqrt{g_i}}\frac{\partial\sqrt{g_i}}{\partial q_i}y_i+\frac{\hbar}{i}\frac{\partial y_i}{\partial q_i}\right]|\psi\rangle$$

$$=-\frac{\hbar}{i}\langle\chi|\sum_{t_i=\bar{t}_1}^{t_i=\bar{t}_2}y_i\frac{\partial\mathscr{F}}{\partial q_i}|\psi\rangle\,. \tag{70}$$

If $\bar{t}_1$ and $\bar{t}_2$ are far apart, we can suppose that, as we have proved in the classical case, if the action is invariant with respect to the transformation $q\to q+ay$, the expression on the left of (70) multiplying $\mathscr{F}$, can be expressed as a difference $I_{\bar{t}_1}-I_{\bar{t}_2}$, where the functional $I_{\bar{t}_2}$ involves the coordinates in the neighbourhood of $\bar{t}_2$ and $I_{\bar{t}_1}$ involves, in the same way, coordinates for times in the neighbourhood of $\bar{t}_1$.

If the expression, $\mathscr{F}$, involves coordinates for only a finite range around some time, $\bar{t}_0$, and $\bar{t}_1$ and $\bar{t}_2$ are outside this range ($\bar{t}_2>\bar{t}_0>\bar{t}_1$) then the right side is independent of $\bar{t}_1$ and $\bar{t}_2$. This is the analogue of the fact that I is a constant of the motion. In this case also, the left side becomes analogous to the quantity $IF-FI$, and the right side is just a differential operation performed on F. The differential operation is characteristic of the group of transformations from which I is derived.

Thus, for a displacement in the x direction, the differential operator is $\frac{\partial}{\partial x}$, the corresponding constant of the motion is momentum in the x direction, as we have the analogue of the operator equation $p_xF-Fp_x=\frac{\hbar}{i}\frac{\partial F}{\partial X}$. More accurately, we have the analogue of the equation, $e^{\frac{i}{\hbar}H(\bar{t}_2-\bar{t}_0)}p_xe^{-\frac{i}{\hbar}H(\bar{t}_2-\bar{t}_0)}F-Fe^{\frac{i}{\hbar}H(\bar{t}_0-\bar{t}_1)}p_xe^{-\frac{i}{\hbar}H(\bar{t}_0-\bar{t}_1)}=\frac{\hbar}{i}\frac{\partial F}{\partial X}$ for all $\bar{t}_1$ and $\bar{t}_2$. For the time-displacement the differential operator is $\frac{d}{dt}-\frac{\partial}{\partial t}$ and the constant is the negative of the energy, so that we have the analogue of $HF-FH=\frac{\hbar}{i}\frac{dF}{dt}-\frac{\hbar}{i}\frac{\partial F}{\partial t}$. If F does not depend explicitly on the time, we can write the right side in the usual way, as $\frac{\hbar}{i}\dot{F}$.

For the energy expression, classically $y(\sigma)=\dot{q}(\sigma)$. The formula (60) can be deduced from (70) most simply by writing for y_i the form, $y_i = \frac{1}{2}\left[\frac{q_{i+1}-q_i}{t_{i+1}-t_i} + \frac{q_i-q_{i-1}}{t_i-t_{i-1}}\right]$ in the case that the coordinates are rectangular. The method described in connection with (60) (page 39) of augmenting all the times between $\bar{t}_1$ and $\bar{t}_2$ by a fixed amount δ may be applied here also to obtain an alternative expression for the energy.

7. The Role of the Wave Function

The problem discussed in this section is that of the existence of a wave function for times between T_1 and T_2.

It is to be noted, that in view of the relation (68) it is no longer possible to express the formula for the averages in some such simple form as $\int \phi_2^* F \phi_1 d\,\text{Vol}$. Suppose that F is especially simple so that it contains only the coordinate q_0 itself. (For example, perhaps we want to compute the average value of q_0.) According to Eq. (68), it may be expressed in the form,

$$\int \rho(q_0) F(q_0) \sqrt{g_0}\, dq_0\,, \tag{71}$$

where $\rho(q_0)$ is the result of integrating the integrand of (68) with respect to every variable q_i except q_0. This is to be compared to the usual expression when a Hamiltonian exists,

$$\int \phi_2^*(q_0) F(q_0) \phi_1(q_0) \sqrt{g_0}\, dq_0\,. \tag{72}$$

They would be equivalent if $\rho(q_0)$ could be expressed as the product of two functions $\phi_2^*(q_0)$ and $\phi_1(q_0)$ in a natural and useful way. In general, however the integral of $\exp \frac{i}{\hbar}\mathscr{A}$ over all the variables except q_0 cannot be so expressed. For the particular case that $\mathscr{A}$ is the integral of an ordinary Lagrangian function of velocity and position, the exponential can be broken up into two factors;

$$\exp \frac{i}{\hbar}\left\{ L\left(\frac{q_0 - q_{-1}}{t_0 - t_1}, q_0\right) \cdot (t_0 - t_{-1}) + L\left(\frac{q_{-1} - q_{-2}}{t_{-1} - t_{-2}}, q_{-1}\right) \right.$$
$$\left. \cdot (t_{-1} - t_{-2}) + \dots \right\}$$

and, (73)

$$\exp \frac{i}{\hbar}\left\{ L\left(\frac{q_1 - q_0}{t_1 - t_0}, q_1\right) \cdot (t_1 - t_0) + L\left(\frac{q_2 - q_1}{t_2 - t_1}, q_2\right) \cdot (t_2 - t_1) + \dots \right\}.$$

These factors contain only the variable q_0 in common, so that when the integrations on the other variables are performed in expression (68), the result remains factorable. The quantity $\phi_1(q_0)$ is the result of integrating the first factor, and is expressed in a form exactly similar to (47). The quantity $\phi_2^*(q_0)$ is expressed in terms of χ^* by an equation like one obtains from (47) by taking the complex conjugate (i.e., (48)).

We can take the viewpoint, then, that the wave function is just a mathematical construction, useful under certain particular conditions to analyze the problem presented by the more generalized quantum mechanical equations (68) and (69), but not generally applicable. It is not unreasonable that it should be impossible to find a quantity like a wave function, which has the property of describing the state of the system at one moment, and from which the state at other moments may be derived. In the more complicated mechanical systems (e.g., the example, (67)) the state of motion of a system at a particular time is not enough to determine in a simple manner the way that the system will change in time. It is also necessary to know the behaviour of the system at other times; information which a wave function is not designed to furnish. An interesting, and at present unsolved, question is whether there exists a quantity analogous to a wave function for these more general systems, and which reduces to the ordinary wave function in the case that the action is the integral of a Lagrangian. That such exist is, of course, not at all necessary. Quantum mechanics can be worked entirely without a wave function, by speaking of matrices and expectation values only. In practice, however, the wave function is a great convenience, and dominates most of our thought in quantum mechanics. For this reason we shall find it especially convenient, in interpreting the physical meaning of the theory, to assume our mechanical systems is such that, no matter how complex between the time T_1 and T_2, outside of this range the action is the integral of a Lagrangian. In this way we may speak of the state of the system at times T_1 and T_2, at least, and represent it by a wave function. This will enable us to describe the meaning of the new generalization in terms with which we are already familiar. This we do in the next section.

8. Transition Probabilities

We shall suppose, as suggested above, that our action has the form of the integral of a Lagrangian for times T_2 or later, and for times T_1 or earlier, but that it is arbitrary in between. In this way, we may speak of the state of the system at time T_1 as being given by a wave function ψ, and of the state of the system at time T_2, by a wave function χ. We can then make the physical assumption that *the probability that the system in state ψ at time T_1 will be found, at the time T_2, in the state χ is the square of the absolute value of the quantity $\langle\chi|1|\psi\rangle$*. The quantity may be defined by the expression (68) with F replaced simply by unity.

We can define other physical quantities in terms of this, by determining the changes in this probability, or rather in the quantity $\langle\chi|1|\psi\rangle$, produced by perturbations of the motion.

We shall indicate by a subscipt the action for which the quantity $\langle\chi|1|\psi\rangle$ is calculated, by writing $\langle\chi|1|\psi\rangle_{\mathscr{A}}$ if the action is $\mathscr{A}$. Suppose the action is slightly altered (in the interval T_1 to T_2) to become $\mathscr{A}+\varepsilon\mathscr{F}$ where ε is a very small parameter. From the form of equation (68) we would have,

$$\langle\chi|1|\psi\rangle_{\mathscr{A}+\varepsilon\mathscr{F}} = \langle\chi|e^{\frac{i\varepsilon}{\hbar}\mathscr{F}}|\psi\rangle_{\mathscr{A}} \tag{74}$$

so that if ε is small enough to insure convergence, we may write,

$$\langle\chi|1|\psi\rangle_{\mathscr{A}+\varepsilon\mathscr{F}} = \langle\chi|1|\psi\rangle_{\mathscr{A}} + \frac{i\varepsilon}{\hbar}\langle\chi|\mathscr{F}|\psi\rangle_{\mathscr{A}} - \frac{\varepsilon^2}{2\hbar^2}\langle\chi|\mathscr{F}^2|\psi\rangle_{\mathscr{A}} = \dots \tag{75}$$

We can therefore interprete $\langle\chi|\mathscr{F}|\psi\rangle_{\mathscr{A}}$ by saying it is $\frac{\hbar}{i}\frac{d}{d\varepsilon}(\langle\chi|1|\psi\rangle_{\mathscr{A}+\varepsilon\mathscr{F}})$ at $\varepsilon = 0$. It should be emphasized here that $\langle\chi|\mathscr{F}\mathscr{G}|\psi\rangle_{\mathscr{A}}$ where $\mathscr{F}$ and $\mathscr{G}$ are any two functionals cannot in general be written in a way analogous to a matrix product (e.g., $\sum_m\langle\chi|\mathscr{F}|\phi_m\rangle\langle\phi_m|\mathscr{G}|\psi\rangle$) as can be done in the usual mechanics. (This is because $\mathscr{F}$ and $\mathscr{G}$ may overlap in time, and so neither be before the other.) The term $\langle\chi|\mathscr{F}\mathscr{G}|\psi\rangle_{\mathscr{A}}$ may either be interpreted as $\langle\chi|\mathscr{F}|\psi\rangle_{\mathscr{A}}$ with $\mathscr{F}$ replaced by $\mathscr{F}\mathscr{G}$, or alternatively, as the first order change in $\langle\chi|\mathscr{F}|\psi\rangle_{\mathscr{A}}$ on changing $\mathscr{A}$ to $\mathscr{A}+\varepsilon\mathscr{F}$ (see equation (76)).

We have, incidently, derived a perturbation equation (75), which may be easily generalized (change $\mathscr{A}$ to $\mathscr{A}+\gamma\mathscr{G}$ in (74), differentiate both sides with respect to γ, and set $\gamma = 0$) to read,

$$\langle\chi|\mathscr{G}|\psi\rangle_{\mathscr{A}+\varepsilon\mathscr{G}} = \langle\chi|\mathscr{G}|\psi\rangle_{\mathscr{A}} + \frac{i\varepsilon}{\hbar}\langle\chi|\mathscr{F}\mathscr{G}|\psi\rangle_{\mathscr{A}} - \frac{\varepsilon^2}{2\hbar^2}\langle\chi|\mathscr{F}^2\mathscr{G}|\psi\rangle_{\mathscr{A}} + \dots \tag{76}$$

This permits us to express the average of a functional for one action function in terms of averages of other functionals for a slightly different action. For some particular problems, such as, for example, the electrodynamic one, it may turn out that the action may be considered as the sum of two terms, $\mathscr{A}_0 + \mathscr{A}_1$, the first expressible as the integral of a Lagrangian, while the second, not so expressible, may be considered as a small perturbation. Equation (76) then permits the actual matrix elements to be expressed in terms of the matrix elements with the Lagrangian action, $\mathscr{A}_0$, alone. Since for this action, $\mathscr{A}_0$, the problem is comparatively simple because wave functions can be defined, the relation (76) will serve as a practical method for solving problems in these cases.

Perturbations may also be considered as producing transitions. Suppose the state of the system at the early time T_1 was ψ. Let us choose some state χ, at the time T_2, and ask for the probability, with the perturbed action $\mathscr{A} + \varepsilon\mathscr{F}$, that the system will be found in this state at this time. It is just $|\langle\chi|\psi\rangle_{\mathscr{A}+\varepsilon\mathscr{F}}|^2$. We shall further suppose that χ is so chosen that were it not for the perturbation, $\varepsilon\mathscr{F}$, the system would have no chance of being found in the state χ;

that is, $\langle\chi|1|\psi\rangle_{\mathscr{A}} = 0$. Therefore from equation (75), to the order ε^2 the probability that the system originally in state ψ, is found in the state χ at that time T_2, due to the perturbation — i.e., the transition probability is just,

$$\frac{1}{\hbar^2}\left|\langle\chi|\varepsilon\mathscr{F}|\psi\rangle_{\mathscr{A}}\right|^2 . \tag{77}$$

For the special case of a simple perturbing potential acting for a time 0 to T, we have $\varepsilon\mathscr{F} = -\int_0^T V dt$, so that our transition probability becomes the more usual expression (Compare, Dirac, "The Principles of Quantum Mechanics", p. 177, (Eq. (20))

$$\frac{1}{\hbar^2}\left|\langle\chi|\int_0^T V dt|\psi\rangle_{\mathscr{A}}\right|^2 .$$

It is of interest to notice that if we are given that the wave function at time T_1 is ψ_{T_1}, although we cannot trace the behaviour of the function through the interval T_1 to T_2 we can, nevertheless, answer the question, "What will the wave function be at the time T_2, and later times?" (Of course, if we know the wave function at the time T_2 we can find it later, since it satisfies a Shrödinger equation from that time on.) If we call the wave function at time T_2, ψ_{T_2}, and expand it in terms of a complete set of orthonormal wave functions χ_n at that time, say $\psi_{T_2} = \sum a_n\chi_n$, the coefficients a_n will be just $\langle\chi_n|1|\psi_{T_1}\rangle_{\mathscr{A}}$. Therefore, we have,

$$\psi_{T_2} = \sum_n \chi_n\langle\chi_n|1|\psi_{T_1}\rangle_{\mathscr{A}} .$$

On account of the form (68) for calculating $\langle\chi_n|1|\psi_{T_1}\rangle_{\mathscr{A}}$ we find that this may also be expressed as,

$$\psi_{T_2}(Q) = \int \exp\left\{\frac{i}{\hbar}\mathscr{A}(Q, q_{m-1}, \ldots, q_0, q_{-1}, \ldots, q_{T_1})\right\}$$
$$\psi_{T_1}(q_{T_1}) \cdot \frac{\sqrt{g}dq_{m-1}\cdots\sqrt{g}dq_{T_1}}{A\cdots A\cdots A} \tag{78}$$

in the limit when the subdivisions in time become infinitely fine, where $\mathscr{A}(Q \cdots q_{T_1})$ is that part of the action functional applicable for times from T_1 to T_2 (compare (47)).

9. Expectation Values for Observables

The physical interpretation which is given in the above section, although the only consistent one available, is rather unsatisfactory. This is because the interpretation requires the concept of states representable by a wave function, while we have pointed out that such a representation is in general impossible. We are therefore forced to alter our mechanical problem so that the action has a simple form at large future and past times, so that we may speak of a wave function at these times, at least. This difficulty also finds itself reflected in the mathematical formulation of the operations performed in calculating average values from the equation (68). We have not defined precisely what is to be done when the action does not become simple at times far from the present.

One possibility that suggests itself is to devise some sort of limiting process so that the interpretation of the last section could be used, and the limit taken as $T_1 \to -\infty$ and $T_2 \to +\infty$. The author has made several attempts in this direction but they all appear artificial, having mathematical, rather than physical, content.

An alternative possibility is to avoid the mention of wave functions altogether, and use, as the fundamental physical concept, the expectation value of a quantity, rather than a transition probability. The work done in this connection, which is presented in this section, is admittedly very incomplete and the results tentative. It is included because many of the formulas derived would seem to be of value, and the author believes that the solution to the problem of physical interpretation will lie somewhere in this direction.

In ordinary quantum mechanics the matrix element of an operator A, between two states ψ_m and ψ_n, is given by,

$$A_{mn} = \int \psi_m^* A \psi_n d\,\mathrm{Vol}\ .$$

The expected value for the quantity represented by the operator A, for the state represented by the wave function ψ_n is, $A_{nn} = \int \psi_n^* A \psi_n d\,\text{Vol}$. In an exactly similar way, we have used our definition (68) for a matrix element of a functional F between the state whose wave function at the time T_2 is χ, and state whose wave function at the time T_1 is ψ. To compute the expected value of the functional whose wave function at the time T_1 is ψ_{T_1}, we calculate $\langle\psi_{T_2}|F|\psi_{T_1}\rangle$, where ψ_{T_2} is given in terms of ψ_{T_1} by the equation (78).

Another important quantity in quantum mechanics is the trace of a matrix,[14] $\text{Tr}[A] = \sum_{all\,n} A_{nn}$. It measures the relative (unnormalized) average expected value when *a priori* each state ψ_n is considered as equally likely. We shall speak of it simply is the expectation of A.

Let us suppose A is an operator which has only certain particular eigenvalues a_n, so that $A\chi_n = a_n\chi_n$, for some set of function χ_n. Let us also suppose $F_{a_n}(x)$ is a function of x which is zero unless $x = a_n$, and $F_{a_n}(a_n) = 1$. Then let us find the trace, $\text{Tr}[BF_{a_n}(A)]$. ($F_{a_n}(A)$ is a projection operator). The matrix $B \cdot F_{a_n}(A)$ has for its k, l element, in a representation with the functions χ_n, $[BF_{a_n}(A)]_{kl} = \sum_m B_{km}[F_{a_n}(A)]_{ml}$. But, since A is diagonal in this representation, $[F_{a_n}(A)]_{ml} = 0$ if $m \neq l$, and equals $F_{a_n}(a_m)$ otherwise. Therefore $[B \cdot F_{a_n}(A)]_{kl} = B_{kl}F_{a_n}(a_l)$. Now, $F_{a_m}(a_l) = 0$, unless $l = m$ so that we find, $\text{Tr}[BF_{a_m}(A)] = B_{mm}$. That is to say, the trace of $BF_{a_m}(A)$ is the expected value of B for the state for which the quantity A has the value a_m.

In a like manner it is not hard to show that, $\text{Tr}[F_{b_n}(B) \cdot F_{a_m}(A)]$ is equal to the probability that the quantity B will be found to have the value b_n in a state where it is known that A has the value a_m (neglecting degeneracies).

These examples are given to remind the reader of the fact that by means of the concept of trace all of the important physical concepts

[14] See J. von Neumann, "Mathematische Grundlage der Quantenmechanik" (1932), p. 93, ff.

can be derived. What corresponds to taking a trace in our form of quantum mechanics?

As we can see from equations (68) and (78) the expression for $\langle\psi_{T_2}|\mathscr{F}|\psi_{T_1}\rangle$ can be written in the form, $\int\rho(q,q')\psi^*_{T_1}(q)\psi_{T_1}(q')dqdq'$ where $\rho(q,q')$ is given by a complicated expression obtained by substituting (78) into (68). That is to say, what corresponds to the diagonal element A_{nn} may be written in our case as $\int\rho(q,q')\psi^*_n(q)\psi_n(q')dqdq'$. The sum of the diagonal elements, and therefore the trace, corresponds to

$$\sum_n\int\rho(q,q')\psi^*_n(q)\psi_n(q')dqdq'\,.$$

As is well known, however, the sum over all n of $\psi^*_n(q)\psi_n(q')$ is just $\delta(q-q')$, so that what corresponds to the trace of $\mathscr{F}$ is $\int\rho(q,q)dq$. We are therefore led to consider the quantity,

$$\begin{aligned}\mathrm{Tr}\langle\mathscr{F}\rangle = \int &\exp\left\{-\frac{i}{\hbar}\mathscr{A}[q_{T_2},q'_m,\ldots,q'_{-m'+1},q_{T_1}]\right\}\\ &\times\exp\left\{\frac{i}{\hbar}\mathscr{A}[q_{T_2},q_m,\ldots,q_{-m'+1},q_{T_1}]\right\}\\ &\times\mathscr{F}(\ldots q_1,q_0\ldots)\cdot\sqrt{g_{T_2}}\,dq_{T_2}\cdot\sqrt{g_{T_1}}\,dq_{T_1}\\ &\times\frac{\sqrt{g}\,dq'_m\cdots\sqrt{g}dq'_{-m'+1}}{A^*\cdots A^*}\cdot\frac{\sqrt{g}dq_m\cdots\sqrt{q}dq_{-m'+1}}{A\cdot A}\,.\end{aligned}\tag{79}$$

The passage to the limit of infinitely fine subdivisions is implied, as usual. Since now there are no wave functions, and therefore nothing special about the times T_1 and T_2, we can consider the true trace to be the limit of the $\mathrm{Tr}\langle\mathscr{F}\rangle$, defined in (79), for a sequence of mechanical systems each of which has an action identical to the true one for successively longer time intervals. (The problem of convergence is ever present.)

The trace defined in (79) is identical to the usual trace of quantum mechanics if $\mathscr{F}$ is a function of one coordinate (e.g., q_0) only, and the action is the time integral of a Lagrangian. It lacks, however, in the general case, one important property, and it is this that makes the results of this section so uncertain. The trace of an arbitrary functional is not always a real number!

We lack some condition on the functionals which we are to place into (79), in order to obtain a real value, so that we can say that the functional represents some real observable quantity, the expectation of which is the trace. That is to say, we lack the condition on a functional that it represent an observable, analogous to the condition in ordinary quantum mechanics that an operator, to represent an observable, must be Hermitian. The correct criterion is not known to the author. The most obvious suggestions revolve around the generalization to (79) which is obtained by letting $\mathscr{F}$ be a function of the q' variables, as well as of q. A real trace is obtained if $\mathscr{F}$ is any function symmetrical with respect to interchange of each q with the corresponding q' (i.e., if $\mathscr{F}(\ldots q_1, q_0, \ldots; \ldots q_1', q_0', \ldots) = \mathscr{F}(\ldots q_1', q_0', \ldots; \ldots q_1', q_0)$. For example $\mathscr{F}$ might be a function of $\frac{1}{2}(q_j + q_j')$ only. This symmetry condition may be all that is necessary to insure that the functional correspond to a real observable. The product, and the sum of two such symmetrical functionals is again symmetrical.

We pass from the general problem of a criterion for any functional to that of trying to determine the form of certain special functionals which we should like to identify with special observables (in particular, projection operators). Let us first try to find the $\mathscr{F}$ that we are to place in (79) so that the resulting trace is the probability that q at the time $\bar{t}_2$ (i.e., $q(\bar{t}_2)$) has the value b if it is known that q at the time $\bar{t}_1$ has the value a. If $\mathscr{A}$ is the integral of a Lagrangian the answer is simply $\mathscr{F} = \delta(q_{\bar{t}_2} - b)\delta(q_{\bar{t}_1} - a)$, as can be immediately verified. On the other hand, I have not succeeded in prove that the trace of this quantity is in general real. The trace of,

$$\delta\left(\frac{q_{\bar{t}_2} + q'_{\bar{t}_2}}{2} - b\right) \cdot \delta\left(\frac{q_{\bar{t}_1} + q'_{\bar{t}_1}}{2} - a\right)$$

is, however, real and gives the same value to the desired probability if the action is an integral of a Lagrangian.

We may therefore tentatively assume that,

$$\mathrm{Tr}\left\langle \delta\left(\frac{q_{\bar{t}_2} + q'_{\bar{t}_2}}{2} - b\right) \cdot \delta\left(\frac{q_{\bar{t}_1} + q'_{\bar{t}_1}}{2} - a\right)\right\rangle \cdot db$$

gives the relative probability that, if q has the value a at the time $\bar{t}_1$, a measurement of q at the time $\bar{t}_2$ will lead to b, in the range db. (The absolute probability may be gotten by dividing by $\mathrm{Tr}\,\langle\delta(\frac{q_{\bar{t}_1}+q'_{\bar{t}_1}}{2} - a)\rangle$).

We shall likewise assume that

$$\mathrm{Tr}\left\langle \delta\left(\frac{1}{\varepsilon}\left[\frac{q_{\bar{t}_2+\varepsilon} + q'_{\bar{t}_2+\varepsilon}}{2} - \frac{q_{\bar{t}_2} + q'_{\bar{t}_2}}{2}\right] - v\right) \cdot \delta\left(\frac{q_{\bar{t}_1} + q'_{\bar{t}_1}}{2} - a\right)\right\rangle dv$$

gives the relative probability that, if q has the value a at the time $\bar{t}_1$, a measurement of velocity at the time $\bar{t}_2$ will lead to v in the range dv. That this gives the correct answer, in the case of a Lagrangian action which involves the coordinate q in the kinetic energy term as $\frac{1}{2}m\dot{q}^2$, is shown below.

In a similar way we can define the probabilities for any quantities involving linear combinations of the coordinates (and that includes velocities, accelerations, etc.). For example, the probability that the difference in the coordinate q at the time $\bar{t}_3$ and its value at the time $\bar{t}_2$ is between b and $b + db$, when it is known that at the time $\bar{t}_1$ the velocity plus c times the position is a, is given by the trace of,

$$\delta\left(\frac{q_{\bar{t}_2} + q'_{\bar{t}_2}}{2} - \frac{q_{\bar{t}_3} + q'_{\bar{t}_3}}{2} - b\right)$$
$$\times\,\delta\left(\frac{q_{\bar{t}_1+\varepsilon} + q'_{\bar{t}_1+\varepsilon}}{2\varepsilon} - \frac{q_{\bar{t}_1} + q'_{\bar{t}_1}}{2\varepsilon} + c(q_{\bar{t}_1} + q'_{\bar{t}_1}) - a\right) \cdot db\,.$$

(This has been checked for a free harmonic oscillator). It is possible that this is even true if $\bar{t}_3$ is before $\bar{t}_1$, and $\bar{t}_2$ after $\bar{t}_1$.

We shall now show that

$$\mathrm{Tr}\left\langle \delta\left(\frac{1}{\varepsilon}\left[\frac{q_{\bar{t}_2+\varepsilon}+q'_{\bar{t}_2+\varepsilon}}{2}-\frac{q_{\bar{t}_2}+q'_{\bar{t}_2}}{2}\right]-v\right)\cdot\mathscr{G}\right\rangle,$$

where $\mathscr{G}$ involves times earlier than t_2 does agree with the usual form for finding the probability of a given momentum, mv, if the action is the integral of a Lagrangian, say $\frac{1}{2}m\dot{q}^2 - V(q)$. In the expression (79) the integral on q_{T_2} can be immediately performed, inasmuch as, (see (45))

$$\int \frac{dq_{T_2}}{A}\cdot e^{\frac{i\varepsilon}{\hbar}\left[\frac{m}{2}\left(\frac{q_{T_2}-q_m}{\varepsilon}\right)^2 - V(q_{T_2})\right]}\cdot e^{-\frac{i\varepsilon}{\hbar}\left[\frac{m}{2}\left(\frac{q_{T_2}-q'_m}{\varepsilon}\right)^2 - V(q_{T_2})\right]}$$
$$= \delta(q_m - q'_m)\cdot A^*\,.$$

Thus the integral on q'_m means merely replacing q'_m by q_m. We have thus again the same expression as (79) with one term integrated from the end. We can repeat this process many times, telescoping the Lagrangians, until we come to the term $q_{t_2}+\varepsilon$. (That is to say, we could have taken T_2 equal to $t_2+\varepsilon$ without loss of generality). Let us suppose also that we have integrated all variables following q_{t_2} and q'_{t_2} and the net result is $\rho(q_{t_2}, q'_{t_2})$. (It can be expressed in this form for any $\mathscr{G}$ satisfying our condition in virtue of the form of the action.) That is to say, we must calculate,

$$\int \delta\left(\frac{1}{\varepsilon}\left[q_{t_2+\varepsilon}-\frac{q_{t_2}+q'_{t_2}}{2}\right]-V\right)$$
$$\times\, e^{-\frac{i\varepsilon}{\hbar}\left[\frac{m}{2}\left(\frac{q_{t_2+\varepsilon}-q'_{t_2}}{\varepsilon}\right)^2 - V(q_{t_2+\varepsilon})\right]}$$
$$\times\, e^{+\frac{i\varepsilon}{\hbar}\left[\frac{m}{2}\left(\frac{q_{t_2+\varepsilon}-q_{t_2}}{\varepsilon}\right)^2 - V(q_{t_2+\varepsilon})\right]}$$
$$\times\, \rho(q_{t'_2}, q_{t_2})\cdot dq_{t_2+\varepsilon}\cdot\frac{dq'_{t_2}}{\sqrt{\frac{-2\pi\varepsilon\hbar i}{m}}}\cdot\frac{dq_{t_2}}{\sqrt{\frac{2\pi\hbar\varepsilon i}{m}}}\,.$$

The phases of the exponentials, when combined, reduce simply to

$$\frac{i\varepsilon}{\hbar}\cdot\frac{m}{2}\left[\left(\frac{q_{t_2+\varepsilon}-q_{t_2}}{\varepsilon}\right)^2-\left(\frac{q_{t_2+\varepsilon}-q'_{t_2}}{\varepsilon}\right)^2\right].$$

The integral on the δ function, over $q_{t_2+\varepsilon}$, requires that one substitute for this quantity $q_{t_2+\varepsilon}=\frac{q_{t_2}+q'_{t_2}}{2}+v\varepsilon$, and multiply by ε, to obtain, in the final result,

$$\int\frac{m}{2\pi\hbar}e^{-\frac{imv}{\hbar}q_{t_2}}\cdot e^{+\frac{imv}{\hbar}q'_{t_2}}\rho(q_{t_2},q'_{t_2})dq_{t_2}dq'_{t_2}\,.$$

This agrees with the usual expression for the probability of a given momentum p, if $p=mv$. (The extra factor of normalization, comes from the fact that $dp=mdv$.)

10. Application to the Forced Harmonic Oscillator

As a special problem, because we shall need the results in the next section, we consider, from the point of view of the modified quantum mechanics, the problem of the forced harmonic oscillator; that is, an oscillator interacting with another system. This problem, when the oscillator is interacting with a Lagrangian system, can of course be handled by the usual methods of quantum mechanics. We shall see, however, that the added power of looking at all the times at once, so to speak, which arises in such equations as (68), has some advantages. With a wave function, the oscillator and the interacting system are so firmly interlocked, mathematically, that it is hard to study the properties of the oscillator without, at the same time, solving for the motion of the interacting system. We shall be able here, however, to solve that half of the problem which involves the oscillator, without solving the entire problem.

If x is the coordinate of our oscillator, the action is of the form,

$$\mathscr{A}=\mathscr{A}_0+\int dt\left\{\frac{m\dot{x}^2}{2}-\frac{m\omega^2x^2}{2}+\gamma(t)x\right\},\tag{80}$$

where $\mathscr{A}_0$ is the action of the other particles of the system of which the oscillator is a part, and $\gamma(t) \cdot x$ is the interaction of the oscillator with the rest of the system. Thus, if we symbolize the coordinates of the rest of the system by Q, $\gamma(t)$ is some functional of Q. (We might also contemplate that the oscillator is not interacting with any other quantum mechanical system, but is simply acted on by a forcing potential. In this case, $\gamma(t)$ is a simple function of t, and represents the force acting on the oscillator at time t.) We shall suppose that $\gamma(t)$ is zero for times outside the range 0 to T, and shall compute matrix elements of the form $\langle \chi_T|1|\psi_0\rangle_{\mathscr{A}}$ where ψ_0 is a wave function at time $t = 0$, (it involves Q as well as x), and χ_T is a wave function at time $t = T$. Writing this in more detail, by equation (68) it is,

$$\langle \chi_T|1|\psi_0\rangle_{\mathscr{A}} = \int \chi_T(Q_m, x_m) \exp \frac{i}{\hbar}\left\{ \mathscr{A}_0[\cdots Q_j \cdots] + \sum_{i=0}^{m-1} \right.$$

$$\left. \times \left[\frac{m}{2}\left(\frac{x_{i+1} - x_i}{t_{i+1} - t_i}\right)^2 - \frac{m\omega^2 x_i^2}{2} + \gamma_i x_i \right] \cdot (t_{i+1} - t_i) \right\}$$

$$\times\, \psi_0(Q_0, x_0) \frac{\sqrt{g} dQ_m \cdots \sqrt{g} dQ_0}{A_m \cdots A_1}$$

$$\times \frac{dx_m \cdot dx_{m-1} \cdots dx_0}{\sqrt{\frac{2\pi i\hbar}{m}(t_m - t_{m-1})} \cdots \sqrt{\frac{2\pi i\hbar}{m}(t_1 - t_0)}}, \tag{81}$$

where the A_m are the normalizing constants appropriate to the action $\mathscr{A}_0[Q_j]$, and Q_i is the variable Q at time t_i, x_i the variable x at the time t_i, and $\gamma_i = \gamma(t_i)$, a function, perhaps, of the Q_i. We are to set $t_m = T$ and $t_0 = 0$.

Inasmuch as the integrand is a quadratic function of the x_i (for $i \neq 0$ or m) we may actually perform the integration over these x_i, and leave the integrations on Q_i to be performed later. We will simplify the work by taking all the intervals, $t_{i+1} - t_i$, equal, and equal to ε. We are therefore led to consider the quantity,

$$G_\gamma(x_m, x_0; T) = \lim_{\substack{\varepsilon\to 0\\ m\varepsilon\to T}} \iint \cdots \int \exp\left\{\frac{i\varepsilon}{\hbar}\sum_{i=0}^{m-1}\left[\frac{m}{2}\left(\frac{x_{i+1}-x_i}{\varepsilon}\right)^2 - \frac{m\omega^2 x_i^2}{2} + \gamma_i x_i\right]\right\}\frac{dx_{m-1}\cdots dx_1}{\sqrt{\frac{2\pi\varepsilon\hbar i}{m}}\cdots\sqrt{\frac{2\pi\varepsilon\hbar i}{m}}}\,. \tag{82}$$

We shall perform the integrations one after the other, starting with x_1, then x_2, etc. We shall determine the result by a recursive method. We can guess, that after the integrations from x_1 to $x_i - 1$ have been performed, the integrand will depend, except for a factor involving other x's, quadratically on the variable x_i, according to the following form;

$$A_i e^{\frac{i}{\hbar}(\alpha_i x_i^2 + \beta_i x_i x_{i+1} + \delta_i x_i + \eta_i)}\,.\,\frac{dx_i}{\sqrt{\frac{2\pi\hbar\varepsilon i}{m}}}\,, \tag{83}$$

where A_i, δ_i, η_i are constants, independent of x_i, x_{i+1}, etc., which are to be found. The integral on x_i may now be performed, by writing the exponent as,

$$\frac{i}{\hbar}\alpha_i\left[x_i + \frac{\beta_i x_{i+1}+\delta}{2\alpha_i}\right]^2 - \frac{i}{\hbar}\frac{(\beta_i x_{i+1}+\delta_i)^2}{4\alpha_i} + \frac{i}{\hbar}\eta_i$$

changing the variable x_i to $x_i + \frac{\beta_i x_{i+1}+\delta_i}{2\alpha_i}$ and using $\int e^{\frac{i}{\hbar}\gamma^2}d\gamma = \sqrt{\hbar\pi i}$. We obtain $\frac{A_i}{\sqrt{2\varepsilon\alpha_i/m}}\exp\frac{i}{\hbar}\big(\eta_i - \frac{(\beta_i x_{i+1}+\delta_i)^2}{4\alpha_i}\big)$. Multiplying this by the term

$$\exp\frac{i}{\hbar}\left(\frac{2mx_{i+1}^2}{2\varepsilon} - \frac{mx_{i+1}x_{i+2}}{\varepsilon} - \frac{m\omega^2\varepsilon}{2}x_{i+1}^2 + \varepsilon\gamma_{i+1}x_{i+1}\right)$$

giving the part of the exponential in (82) depending on x_{i+1}, we find that after the integration on x_i is performed the dependence of the integral on x_{i+1} is,

$$\frac{A_i}{\sqrt{2\varepsilon\alpha_i/m}}\exp\frac{i}{\hbar}\left\{\frac{2mx_{i+1}^2}{2\varepsilon}-\frac{mx_{i+1}x_{i+2}}{\varepsilon}-\frac{m\omega^2\varepsilon}{2}x_{i+1}^2\right.$$

$$\left.+\gamma_{i+1}\cdot\varepsilon x_{i+1}+\eta_i-\frac{\beta_i^2x_{i+1}^2}{4\alpha_i}-\frac{\beta_i\delta_i}{2\alpha_i}x_{i+1}-\frac{\delta_i^2}{4\alpha_i}\right\}\cdot\frac{dx_{i+1}}{\sqrt{\frac{2\pi\varepsilon\hbar i}{m}}}.$$

This is again of the form (83), so that our guess is self-consistent if we set,

$$A_{i+1}=\sqrt{\frac{m}{2\varepsilon\alpha_i}}\cdot A_i\,, \tag{84}$$

$$\alpha_{i+1}=\frac{m}{\varepsilon}-\frac{\beta_i^2}{4\alpha_i}-\frac{m\omega^2\varepsilon}{2}\,, \tag{85}$$

$$\beta_{i+1}=-\frac{m}{\varepsilon}\,, \tag{86}$$

$$\delta_{i+1}=\varepsilon\gamma_{i+1}-\frac{2\delta_i\beta_i}{4\alpha_i}\,, \tag{87}$$

$$\eta_{i+1}=\eta_i-\frac{\delta_i^2}{4\alpha_i}\,. \tag{88}$$

We note, therefore, β_i is a constant $-\frac{m}{\varepsilon}$. We shall solve the other equations in the limit $\varepsilon\to 0$ under the assumption (which will be self-consistent, and is therefore correct) that $\alpha_i-\frac{m}{2\varepsilon}$, δ_i, η_i, A_i are all finite.

Replacing (86) in (85) and setting

$$\alpha_i=\frac{m}{2\varepsilon}+\lambda_i\,, \tag{89}$$

find,

$$\lambda_{i+1}=\frac{m}{2\varepsilon}-\frac{m^2}{4\varepsilon^2\left(\frac{m}{2\varepsilon}+\lambda_i\right)}-\frac{m\omega^2\varepsilon}{2}\,.$$

Expanding the fraction $\frac{1}{1+\frac{2\varepsilon}{m}\lambda_i}$ by the series $1-\frac{2\varepsilon}{m}\lambda_i+\frac{4\varepsilon^2}{m^2}\lambda_i^2$ and keeping no further terms, find, $\lambda_{i+1}-\lambda_i=-\frac{2\varepsilon}{m}\lambda_i^2-\frac{m\omega^2}{2}\varepsilon$. As $\varepsilon\to 0$, then, λ_i may be considered as a function of t_i, so that dividing both sides by ε, in the limit we may write,

$$\frac{d\lambda}{dt} = -\frac{2}{m}\lambda^2 - \frac{m\omega^2}{2}\,. \tag{90}$$

This has the solution

$$\lambda = \frac{m\omega}{2} \cot \omega(t + \text{const.})\,. \tag{91}$$

Since, for small t, (e.g., $t = \varepsilon$), α is $\frac{m}{\varepsilon}$, λ must approach $\frac{m}{2\varepsilon}$. This it does if the const. is zero in (91). Therefore,

$$\alpha = \frac{m}{2\varepsilon} + \lambda = \frac{m}{2\varepsilon} + \frac{m\omega}{2} \cot \omega t\,. \tag{92}$$

Place this in (84), yielding, $A_{i+1} = \frac{A_i}{\sqrt{1+\varepsilon\omega \cot \omega t}} \cong A_i\bigl(1 - \frac{\varepsilon\omega}{2} \cot \omega t\bigr)$ for small ε. In the limit then, $\frac{dA}{dt} = -A \cdot \frac{\omega}{2} \cot \omega t$. This implies $A = \frac{\text{const.}}{\sqrt{\sin \omega t}}$. Since for t of the order ε, A is $\frac{i}{\sqrt{\frac{2\pi\varepsilon\hbar i}{m}}}$, the constant must be $\sqrt{\frac{m\omega}{2\pi i\hbar}}$ so that we find,

$$A = \sqrt{\frac{m\omega}{2\pi i\hbar \sin \omega t}}\,. \tag{93}$$

Putting our values of α and β into (87), find,

$$\delta_{i+1} = \varepsilon\gamma_{i+1} + \frac{m\delta_i}{2\varepsilon\bigl(\frac{m}{2\varepsilon} + \frac{m\omega}{2} \cot \omega t\bigr)}\,,$$

which leads to the differential equation $\frac{d\delta}{dt} = \gamma - \delta \cdot \omega \cot \omega t$. This equation has the general solution

$$\delta = \frac{1}{\sin \omega t} \int_0^t \gamma(s) \sin \omega s ds + \frac{\text{const.}}{\sin \omega t}\,.$$

Since, for small values of t, of order ε, δ approaches $-\frac{m}{\varepsilon}x_0$, constant here is $-m\omega x_0$. Therefore,

$$\delta = -\frac{m\omega x_0}{\sin \omega t} + \frac{1}{\sin \omega t} \int_0^t \gamma(s) \sin \omega s ds\,. \tag{94}$$

Replacing α by its leading term, $\frac{m}{2\varepsilon}$, in Eq. (89) we obtain, in the limit, the equation, $\frac{d\eta}{dt} = -\frac{\delta^2}{2m}$. Thus, we find,

$$\eta = -\int^t \frac{[\delta(t)]^2}{2m} dt \tag{95}$$

and is subject to the condition that as $t \to \varepsilon$, $\eta \to \frac{m}{2\varepsilon} x_0^2$. We may now utilize these results to compute $G_\gamma(x_m, x_0; T)$.

Since it is clear from its mode of formation that,

$$e^{\frac{i}{\hbar}\left(\frac{m}{2\varepsilon} x_m^2 - m\frac{x_m x_{m+1}}{\varepsilon}\right)} G_\gamma(x_m, x_0; T)$$
$$= A_m e^{\frac{i}{\hbar}(\alpha_m x_m^2 + \beta_m x_m x_{m+1} + \delta_m x_m + \eta_m)}$$

we may calculate G in the limit as $\varepsilon \to 0$, $m\varepsilon \to T$, from our expressions (92), (86), (94), (95) and (93). After a little algebraical rearrangement, it may be written in the rather convenient form (we have replaced x_m by x, and x_0 by x'),

$$G_\gamma(x, x'; T) = G_0(x - a, x' - b; T) \cdot \exp \frac{i}{2m\omega\hbar}$$
$$\times \left\{ \int_0^T \int_0^t \gamma(s)\gamma(t) \sin \omega(t - s) ds dt + m^2\omega^2 \sin \omega T \cdot a \cdot b \right\}, \tag{96}$$

where

$$a = \frac{1}{m\omega \sin \omega T} \int_0^T \gamma(t) \cos \omega t dt \tag{97}$$

and

$$b = \frac{1}{m\omega \sin \omega T} \int_0^T \gamma(t) \cos \omega(T - t) dt \tag{98}$$

and $G_0(y, y'; T)$ is the value to which $G_\gamma(y, y'; T)$ reduces when $\gamma = 0$, and is the well known generating function for the unperturbed harmonic oscillator;

$$G_0(y,y';T) = \sqrt{\frac{m\omega}{2\pi i\hbar \sin \omega t}} \exp \frac{m\omega i}{2\hbar \sin \omega T} \times \{(y^2 + (y')^2) \cos \omega T - 2yy'\} . \tag{99}$$

Going back to equation (81), we may now express the average in question by the simplified expression,

$$\langle \chi_T | 1 | \psi_0 \rangle_{\mathscr{A}} = \int \chi_T(Q_m, x) e^{\frac{i}{\hbar}\mathscr{A}_0[\ldots Q_i \ldots]} \times G_\gamma(x, x'; T) \psi_0(Q_0, x') dx dx' \cdot \frac{\sqrt{g} dQ_m \cdots \sqrt{g} dQ_0}{A_m \cdots A_1} . \tag{100}$$

We shall need this formula in the next section. It is the analogue, in a sense, of the solution (26), (27), (28) of the equations of motion for the oscillator, in the classical case. Here, as there, the solution for the motion of the oscillator is expressed in terms of the interacting system, without it actually being necessary to solve for the motion of this interacting system.

11. Particles Interacting through an Intermediate Oscillator

The problem which we discuss in this section is the quantum analogue of the problem discussed in Section 4 of the first part of the paper. Given two atoms A and B, each of which interacts with an oscillator O, to which extent can the motion of the oscillator be disregarded and the atoms be considered as interacting directly? This problem has been solved in a special case by Fermi[15] who has shown that the oscillators of the electromagnetic field which represent longitudinal waves could be eliminated from the Hamiltonian, provided an additional term be added representing instantaneous Coulomb

[15] E. Fermi, *Rev. Mod. Phys.* **4**, (1932), p. 131.

interactions between the particles. Our problem is analogous to his except that in the general case, as we can see from the classical analogue, we shall expect that the interactions will not be instantaneous, and hence not expressible in Hamiltonian form.

Drawing on the classical analogue we shall expect that the system with the oscillator is not equivalent to the system without the oscillator for all possible motions of the oscillator, but only for those for which some property (e.g., the initial and final position) of the oscillator is fixed. These properties, in the cases discussed, are not properties of the system at just one time, so we will not expect to find the equivalence simply by specifying the state of the oscillator at a certain time, by means of a particular wave function. It is for this reason that the ordinary methods of quantum mechanics do not suffice to solve this problem.

The natural question to ask, working from analogy with the classical system, is: "What is the expected value of $\mathscr{F}$, a functional involving the particles only, if it is known that at time 0, the position of the oscillator was $x(0) = \alpha$, and that at the time T it was $x(T) = \beta$?" If we can show that the answer to this question is, with a given fixed α and β, the same as the expectation of $\mathscr{F}$ calculated by a formula of exactly the form (79) for an action principle involving the particles alone, then we shall have found the conditions under which a direct interaction can be represented as acting through an intermediate oscillator.

That is, we should like to satisfy, (see p. 53ff),

$$\operatorname{Tr}\left\langle \mathscr{F}\cdot\delta\left(\frac{x_T+x'_T}{2}-\beta\right)\cdot\delta\left(\frac{x_0+x'_0}{2}-\alpha\right)\right\rangle_{A,B,\&O}$$
$$= \text{const.}\cdot\operatorname{Tr}\langle\mathscr{F}\rangle_{A,B}\,, \tag{101}$$

where the trace on the left is computed for the action of the particles and the oscillator, and that on the right only involves the particles. (The constant appears because we are interested only in relative expectations, and can normalize the trace later.)

We shall simplify matters by supposing the particles are represented by a set of coordinates, which we symbolize by Q, that the action for the particles is $\mathscr{A}_0$, that the oscillator has the coordinate $x(t)$, with the Lagrangian $\frac{m}{2}\dot{x}^2 - \frac{m}{2}\omega^2 x^2$, and that the Lagrangian of the interaction is $\gamma(t)x(t)$ where $\gamma(t)$ is some functional of $Q(t)$. Let the action for the particles with the oscillator eliminated be $\mathscr{A}_0 + \mathscr{I}$, where $\mathscr{I}$, a functional of Q only, the action of interaction, is to be found to satisfy (101).

If (101) is to be satisfied for all arbitrary functionals, $\mathscr{F}$, we must have, on account of (79),

$$\begin{aligned}
&\int \exp \frac{i}{\hbar}\left[\mathscr{A}_0 + \int_0^T \gamma(t)x(t)dt + \int_0^T \frac{m}{2}(\dot{x}^2 - \omega^2 x^2)dt\right] \\
&\quad \times \exp -\frac{i}{\hbar}\left[\mathscr{A}_0' + \int_0^T \gamma'(t)x'(t)dt + \int_0^T \frac{m}{2}(\dot{x}'^2 - \omega^2 x'^2)dt\right] \\
&\quad \times \delta\left(\frac{x_T + x_T'}{2} - \beta\right)\delta\left(\frac{x_0 + x_0'}{2} - \alpha\right)dx_T dx_0 \\
&\quad \times \frac{dx\cdots dx}{A\cdots A}\cdot\frac{dx'\cdots dx'}{A^*\cdots A^*} \\
&= \text{const.}\int \exp\frac{i}{\hbar}[\mathscr{A}_0 + \mathscr{I}]\cdot\exp -\frac{i}{\hbar}[\mathscr{A}_0' + \mathscr{I}'] \\
&\quad \times \delta\left(\frac{x_T + x_{T'}'}{2} - \beta\right)\delta\left(\frac{x_0 + x_0'}{2} - \alpha\right)dx_T dx_0 \frac{dx\cdots dx\,dx'\cdots dx'}{A\cdots A^*\cdots},
\end{aligned} \tag{102}$$

where $\mathscr{A}_0'$, γ', and $\mathscr{I}'$ are the same functionals but of a different variable (namely, Q') than $\mathscr{A}$, γ, and $\mathscr{I}$, and the integrations over the x's are to be performed as prescribed in detail in (79).[16] We can divide out the $e^{\frac{i}{\hbar}(\mathscr{A}_0 - \mathscr{A}_0')}$ from each side, of course. Our main problem is to show that the left side, when integrated over all the coordinates of the oscillator, can be expressed as a product of an exponential function of Q' only with an exponential of minus the

[16] Editor's note: The right-hand side of (102) is missing in the original typescript of the thesis. Its form has been inferred by the editor from the surrounding text.

same function of Q only. This will be by no means generally true, and is a consequence of our special choice of δ functions on the left of (102).

Since as far as the variables x are concerned, the action is Lagrangian, the integration over all x and x' from x_{T_2} to x_T can be immediately performed in the way indicated at the end of Section 9 (see page 55). In a like manner the integrations from x_{T_1} to x_0 may be performed. The integrations on the x-primes from $x'_{0+\varepsilon}$ to $x'_{T-\varepsilon}$ can next be performed according to the method of the last section, and they yield $G^*_{\gamma'}(x_T, x_0; T)$. Those on the corresponding intermediate x's result in $G_\gamma(x_T, x_0; T)$. Thus we must prove that,

$$\iint \delta(x_T - \beta)\delta(x_0 - \alpha)G^*_{\gamma'}(x_T, x_0; T)G_\gamma(x_T, x_0; T)dx_T dx_0$$
$$= \text{const.}\, e^{-\frac{i}{\hbar}\mathscr{I}'} e^{\frac{i}{\hbar}\mathscr{I}}\,. \tag{103}$$

The left side is simply $G_{\gamma'}(\beta, \alpha; T)G_\gamma(\beta, \alpha; T)$, and by substituting β, α in the formulas (96)–(99) we can find that this left side is of the form of the right side, if we take const. $= \frac{m\omega}{2\pi\hbar \sin \omega T}$ and $\mathscr{I}$ equal to

$$\mathscr{I} = \frac{m\omega \cot \omega T}{2}[(\beta - b)^2 + (\alpha - a)^2] - \frac{m\omega}{\sin \omega T}(\beta - b)(\alpha - a)$$
$$+ \frac{1}{2m\omega}\int_0^T \int_0^t \gamma(t)\gamma(s) \sin \omega(t - s) dt ds + \frac{m\omega}{2} \sin \omega T \cdot ab$$

with a and b as in (97) and (98). $\mathscr{I}'$ is then the corresponding thing with γ' replacing γ everywhere. After a little rearranging, we get

$$\mathscr{I} = \int_0^T \frac{\alpha \sin \omega(T - t) + \beta \sin \omega t}{\sin \omega T}\gamma(t)dt$$
$$- \frac{1}{m\omega \sin \omega T}\int_0^T \int_0^t \sin \omega(T - t) \sin \omega s \gamma(s)\gamma(t) ds dt$$
$$+ \frac{m\omega \cot \omega T}{2}[\beta^2 + \alpha^2] - \frac{m\omega}{\sin \omega T}\beta\alpha\,. \tag{104}$$

This is identical to the expression (35) obtained for the action in the corresponding classical problem in which the initial value of x is held at $x(0) = \alpha$, and the final value of $x(T) = \beta$. The added constant term has, of course, no significance. (It will cancel a corresponding term in $\mathscr{I}'$.)

Thus, the particles with the action of interaction $\mathscr{I}$, may be replaced by a system with an intermediate oscillator, provided that, in calculating the expectation of any functional of the particles, it is calculated under the conditions that the oscillator's initial position is known to be α and its final position is known to be β. It is to be noted that we have not proved that, in general, the system with the oscillator is equivalent to one without, for that is not true. The equivalence only holds if the oscillator is known to satisfy certain condition.

For the other example, (28), of conditions leading to an action principle, which we have considered in the classical case, we are led, here, to ask the question: "What is the expected value of $\mathscr{F}$ if it is known that for the oscillator,

$$\frac{1}{2}\left[x(0) + x(T)\cos\omega T - \dot{x}(T)\,\frac{\sin\omega T}{\omega}\right] = R_0$$

and that, $\frac{1}{2}\left[x(T) + x(0)\cos\omega T + \dot{x}(0)\frac{\sin\omega T}{\omega}\right] = R_T$?"

To answer this question we must try to satisfy an equation analogous to (102) but with $\delta\left(\frac{x_T + x'_T}{2} - \beta\right) \cdot \delta\left(\frac{x_0 + x'_0}{2} - \alpha\right)$ replaced by,

$$\begin{aligned}
&\delta\Bigg(\frac{1}{4}\Bigg\{x_0 + x'_0 + (x_T + x'_T)\cos\omega T \\
&\qquad - \frac{(x_{T+\varepsilon} + x'_{T+\varepsilon}) - (x'_T + x_T)}{\varepsilon\omega}\sin\omega T\Bigg\} - R_0\Bigg) \\
&\times\delta\Bigg(\frac{1}{4}\Bigg\{x_T + x'_T + (x_0 + x'_0)\cos\omega T \\
&\qquad + \frac{(x_0 + x'_0) - (x'_{-\varepsilon} + x_{-\varepsilon})}{\varepsilon\omega}\sin\omega T\Bigg\} - R_T\Bigg). \qquad (105)
\end{aligned}$$

In this case the integrations on x and x' can only be performed from x_{T_2} to $x_{T+\varepsilon}$, leaving an extra Lagrangian factor, and those from x_{T_1} can proceed only to $x_{-\varepsilon}$. The integrations between x_0 and x_T

can be performed as before, giving G functions. The result is that we must find an $\mathscr{I}$ to satisfy:

$$\int \Bigg\{ dx_{T+\varepsilon} \cdot e^{-\frac{i\varepsilon}{\hbar}\left[\frac{m}{2}\left(\frac{x_{T+\varepsilon}-x'_T}{\varepsilon}\right)^2 - \frac{m\omega^2}{2}x^2_{T+\varepsilon}\right]} \cdot \frac{dx'_T}{\sqrt{\frac{2\pi i\varepsilon\hbar}{-m}}} \cdot G^*_{\gamma'}(x'_T, x'_0; T)$$

$$\cdot \frac{dx'_0}{\sqrt{\frac{2\pi i\varepsilon\hbar}{-m}}} \cdot e^{-\frac{i\varepsilon}{\hbar}\left[\frac{m}{2}\left(\frac{x'_0-x_{-\varepsilon}}{\varepsilon}\right)^2 - \frac{m\omega^2}{2}(x'_0)^2\right]} \cdot \delta\Bigg(\frac{1}{4}\Bigg\{x_0 + x'_0 + (x_T + x'_T)$$

$$\times \cos\omega T - \frac{2x_{T+\varepsilon} - (x'_T + x_T)}{\varepsilon\omega}\sin\omega T\Bigg\} - R_0\Bigg) \cdot \delta\Bigg(\frac{1}{4}\Bigg\{x_T + x'_T$$

$$+ (x_0 + x'_0)\cos\omega T + \frac{(x'_0 + x_0) - 2x_{-\varepsilon}}{\varepsilon\omega}\sin\omega T\Bigg\} - R_T\Bigg)$$

$$\cdot e^{\frac{i\varepsilon}{\hbar}\left[\frac{m}{2}\left(\frac{x_{T+\varepsilon}-x_T}{\varepsilon}\right)^2 - \frac{m\omega^2}{2}x^2_{T+\varepsilon}\right]} \frac{dx_T}{\sqrt{\frac{2\pi i\varepsilon\hbar}{m}}} \cdot G_\gamma(x_T, x_0; T) \cdot \frac{dx_0}{\sqrt{\frac{2\pi i\varepsilon\hbar}{m}}}$$

$$\cdot e^{+\frac{i\varepsilon}{\hbar}\left[\frac{m}{2}\left(\frac{x_0-x_{-\varepsilon}}{\varepsilon}\right)^2 - \frac{m\omega^2}{2}x_0^2\right]} \cdot dx_{-\varepsilon}\Bigg\} = \text{const.}\, e^{-\frac{i}{\hbar}\mathscr{I}'} e^{\frac{i}{\hbar}\mathscr{I}}\,. \tag{106}$$

Since the integration of this complicated expression is perfectly straightforward, we shall not include it here. (It is best to integrate with respect to $dx_{T+\varepsilon}$ and $dx_{-\varepsilon}$ first. The term $e^{+\frac{i\varepsilon}{\hbar}\frac{\omega^2}{2}(x'_0)^2} e^{-\frac{i\varepsilon}{\hbar}\frac{\omega^2}{2}x_0^2}$ can be neglected because it is of no importance in the limit as $\varepsilon \to 0$.) The result is that the left side may be made equal to the right side by choosing the constant as $\frac{2\pi\hbar m\omega}{\sin\omega T}$ and taking $\mathscr{I}$ to be

$$\mathscr{I} = \int_0^T \frac{R_0 \sin\omega(T-t) + R_T \sin\omega t}{\sin\omega T}\gamma(t)dt$$
$$+ \frac{1}{2m\omega}\int_0^T \int_0^t \sin\omega(t-s)\gamma(t)\gamma(s)dtds\,, \tag{107}$$

with a similar expression for $\mathscr{I}'$, obtained by replacing γ by γ'. This is again in agreement with the classical result.

To answer the question: "What is the expectation of $\mathscr{F}$ if it is known that initially the position of the oscillator is w, and its velocity is v?", we must try to satisfy an equation analogous to (102), but with $\delta(\frac{X_T+x'_T}{2}-\beta)\cdot\delta(\frac{X_0+x'_0}{2}-\alpha)$ replaced by

$$\delta\left(\frac{x_0+x'_0}{2}-w\right)\cdot\delta\left(\frac{x_0+x'_0}{2\varepsilon}-\frac{x_{-\varepsilon}+x'_{-\varepsilon}}{2\varepsilon}-v\right).$$

If this is done, however, and the integrations are carried out, the left side of (102) becomes

$$\frac{m}{2\pi\hbar}\exp\frac{i}{2\hbar m\omega}\left\{\int_0^T 2(\gamma(t)-\gamma'(t))(mv\sin\omega t+\omega w\cos\omega t)dt\right.$$
$$\left.+\int_0^T\int_0^t(\gamma(t)-\gamma'(t))(\gamma(s)+\gamma'(s))\sin\omega(t-s)dt\right\}. \quad (108)$$

It is seen that (102) cannot be satisfied by any choice of $\mathscr{I}$, and there now appear in the exponential (108) cross terms between γ and γ', such as $\gamma(t)\gamma'(s)$. This corresponds to the classical finding, that no action exists in case the initial position and velocity are held constant.

These results serve as a confirmation of our formal generalization to systems without a Hamiltonian. They have obvious application to electrodynamics into which we will, however, not go here.

We should like to make a remark, before closing this section, about equation (108). Even though it does not lead to a system which can be expressed by a quantum mechanical least action principle, it is nevertheless correct, of course, that to find the average of the functional $\mathscr{F}$ we must multiply $\mathscr{F}$ by the expression (108) times $\exp\frac{i}{\hbar}(\mathscr{A}_0-\mathscr{A}'_0)$ and integrate over all the Q and Q'. That is to say, the expected value of $\mathscr{F}$ for this system is obtained in a way analogous to (79), except that the phase of the exponential, which in (79) is of the form $\frac{i}{\hbar}\{\mathscr{A}[Q]-\mathscr{A}[Q']\}$ is now of the form $\frac{i}{\hbar}\mathscr{B}(Q,Q')$, where the quantity $\mathscr{B}$ involves the expression in (108).

What should be the behaviour of the system, described in this way, in the classical limit as $\hbar\to 0$? Dirac's argument (see page 27) in this case leads to the conclusion that only those values of Q, and of Q', will be of importance which satisfy both,

$$\frac{\delta\mathscr{B}(Q,Q')}{\delta Q(t)} = 0 \quad \text{and} \quad \frac{\delta\mathscr{B}(Q,Q')}{\delta Q'(t)} = 0\,.$$

Inasmuch as $\mathscr{B}(Q,Q') = -\mathscr{B}(Q',Q)$, the second of these equations results from the first by interchange of the Q and Q'. Therefore one solution of these equations would have the property that $Q(t) = Q'(t)$ where $Q(t)$ satisfies,

$$\left.\frac{\delta\mathscr{B}(Q,Q')}{\delta Q(t)}\right|_{Q'(t)=Q(t)} = 0\,. \tag{109}$$

With $\mathscr{B}$ as given by (108), this leads immediately to the classical equation got by substituting for $x(t)$ from (26) in (24), and replacing $x(0)$ by w, and $\dot{x}(0)$ by v. This suggests a way of quantizing systems which classically do not satisfy a simple principle of least action, but we shall not investigate this here.

12. Conclusion

We have presented, in the foregoing pages, a generalization of quantum mechanics applicable to a system whose classical analogue is described by a principle of least action. It is important to emphasize, however, some of the difficulties and limitations of the description presented here.

One of the most important limitations has already been discussed. The interpretation of the formulas from the physical point of view is rather unsatisfactory. The interpretation in terms of the concept of transition probability requires our altering the mechanical system, and our speaking of states of the system at times very far from the present. The interpretation in terms of expectations, which avoids this difficulty, is incomplete, since the criterion that a functional represent a real physical observable is lacking. It is possible that an analysis of the theory of measurements is required here. A concept such as the "reduction of the wave packet" is not directly applicable, for in the mathematics we must describe the system for all times, and if a measurement is going to be made in the interval of interest, this fact must be put somehow into the equations from the start. Summarizing: a physical interpretation should be sought which does not refer to the behaviour of the system at times very far distant from a present time of interest.

A point of vagueness is the normalization factor, A. No rule has been given to determine it for a given action expression. This question is related to the difficult mathematical question as to the conditions under which the limiting process of subdividing the time scale, required by equations such as (68), actually converges.

The problem of the form that relativistic quantum mechanics, and the Dirac equation, take from this point of view remains unsolved. Attempts to substitute, for the action, the classical relativistic form (integral of proper time) have met with difficulties associated with the fact that the square root involved becomes imaginary for certain values of the coordinates over which the action is integrated.

The final test of any physical theory lies, of course, in experiment. No comparison to experiment has been made in the paper. The author hopes to apply these methods to quantum electrodynamics. It is only out of some such direct application that an experimental comparison can be made.

The author would like to express his gratitude to Professor John A. Wheeler for his continued advice and encouragement.

SPACE-TIME APPROACH TO NON-RELATIVISTIC QUANTUM MECHANICS*

R. P. FEYNMAN

Cornell University, Ithaca, New York

Abstract

Non-relativistic quantum mechanics is formulated here in a different way. It is, however, mathematically equivalent to the familiar formulation. In quantum mechanics the probability of an event which can happen in several different ways is the absolute square of a sum of complex contributions, one from each alternative way. The probability that a particle will be found to have a path $x(t)$ lying somewhere within a region of space-time is the square of a sum of contributions, one from each path in the region. The contribution from a single path is postulated to be an exponential whose (imaginary) phase is the classical action (in units of $\hbar$) for the path in question. The total contribution from all paths reaching x, t from the past is the wave function $\psi(x, t)$. This is shown to satisfy Schroedinger's equation. The relation to matrix and operator algebra is discussed. Applications are indicated, in particular to eliminate the coordinates of the field oscillators from the equations of quantum electrodynamics.

1. Introduction

It is a curious historical fact that modern quantum mechanics began with two quite different mathematical formulations: the differential equation of Schroedinger, and the matrix algebra of Heisenberg. The two, apparently dissimilar approaches, were proved to be mathematically equivalent. These two points of view were destined to complement one another and to be ultimately synthesized in Dirac's transformation theory.

This paper will describe what is essentially a third formulation of non-relativistic quantum theory. This formulation was suggested by some of

* This article was first published in *Rev. Mod. Phys.* **20**, 367–387 (1948).

Dirac's[1,2] remarks concerning the relation of classical action[3] to quantum mechanics. A probability amplitude is associated with an entire motion of a particle as a function of time, rather than simply with a position of the particle at a particular time.

The formulation is mathematically equivalent to the more usual formulations. There are, therefore, no fundamentally new results. However, there is a pleasure in recognizing old things from a new point of view. Also, there are problems for which the new point of view offers a distinct advantage. For example, if two systems A and B interact, the coordinates of one of the systems, say B, may be eliminated from the equations describing the motion of A. The interaction with B is represented by a change in the formula for the probability amplitude associated with a motion of A. It is analogous to the classical situation in which the effect of B can be represented by a change in the equations of motion of A (by the introduction of terms representing forces acting on A). In this way the coordinates of the transverse, as well as of the longitudinal field oscillators, may be eliminated from the equations of quantum electrodynamics.

In addition, there is always the hope that the new point of view will inspire an idea for the modification of present theories, a modification necessary to encompass present experiments.

We first discuss the general concept of the superposition of probability amplitudes in quantum mechanics. We then show how this concept can be directly extended to define a probability amplitude for any motion or path (position *vs.* time) in space-time. The ordinary quantum mechanics is shown to result from the postulate that this probability amplitude has a phase proportional to the action, computed classically, for this path. This is true when the action is the time integral of a quadratic function of velocity. The relation to matrix and operator algebra is discussed in a way that stays as close to the language of the new formulation as possible. There is no practical advantage to this, but the formulae are very suggestive if a gen-

---

1 P. A. M. Dirac, *The Principles of Quantum Mechanics* (The Clarendon Press, Oxford, 1935), second edition, Sec. 33; also, *Physik. Zeits. Sowjetunion* **3**, 64 (1933).

2 P. A. M. Dirac, *Rev. Mod. Phys.* **17**, 195 (1945).

3 Throughout this paper the term "action" will be used for the time integral of the Lagrangian along a path. When this path is the one actually taken by a particle, moving classically, the integral should more properly be called Hamilton's first principle function.

eralization to a wider class of action functionals is contemplated. Finally, we discuss applications of the formulation. As a particular illustration, we show how the coordinates of a harmonic oscillator may be eliminated from the equations of motion of a system with which it interacts. This can be extended directly for application to quantum electrodynamics. A formal extension which includes the effects of spin and relativity is described.

2. The Superposition of Probability Amplitudes

The formulation to be presented contains as its essential idea the concept of a probability amplitude associated with a completely specified motion as a function of time. It is, therefore, worthwhile to review in detail the quantum-mechanical concept of the superposition of probability amplitudes. We shall examine the essential changes in physical outlook required by the transition from classical to quantum physics.

For this purpose, consider an imaginary experiment in which we can make three measurements successive in time: first of a quantity A, then of B, and then of C. There is really no need for these to be of different quantities, and it will do just as well if the example of three successive position measurements is kept in mind. Suppose that a is one of a number of possible results which could come from measurement A, b is a result that could arise from B, and c is a result possible from the third measurement C.[4] We shall assume that the measurements A, B, and C are the type of measurements that completely specify a state in the quantum-mechanical case. That is, for example, the state for which B has the value b is not degenerate.

It is well known that quantum mechanics deals with probabilities, but naturally this is not the whole picture. In order to exhibit, even more clearly, the relationship between classical and quantum theory, we could suppose that classically we are also dealing with probabilities but that all probabilities either are zero or one. A better alternative is to imagine in the classical case that the probabilities are in the sense of classical

[4] For our discussion it is not important that certain values of a, b, or c might be excluded by quantum mechanics but not by classical mechanics. For simplicity, assume the values are the same for both but that the probability of certain values may be zero.

statistical mechanics (where, possibly, internal coordinates are not completely specified).

We define P_{ab} as the probability that if measurement A gave the result a, then measurement B will give the result b. Similarly, P_{bc} is the probability that if measurement B gives the result b, then measurement C gives c. Further, let P_{ac} be the chance that if A gives a, then C gives c. Finally, denote by P_{abc} the probability of all three, i.e., if A gives a, then B gives b, and C gives c. If the events between a and b are independent of those between b and c, then

$$P_{abc} = P_{ab} P_{bc} \,. \tag{1}$$

This is true according to quantum mechanics when the statement that B is b is a complete specification of the state.

In any event, we expect the relation

$$P_{ac} = \sum_b P_{abc} \,. \tag{2}$$

This is because, if initially measurement A gives a and the system is later found to give the result c to measurement C, the quantity B must have had some value at the time intermediate to A and C. The probability that it was b is P_{abc}. We sum, or integrate, over all the mutually exclusive alternatives for b (symbolized by Σ_b).

Now, the essential difference between classical and quantum physics lies in Eq. (2). In classical mechanics it is always true. In quantum mechanics it is often false. We shall denote the quantum-mechanical probability that a measurement of C results in c when it follows a measurement of A giving a by P^q_{ac}. Equation (2) is replaced in quantum mechanics by this remarkable law:[5] There exist complex numbers φ_{ab}, φ_{bc}, φ_{ac} such that

$$P_{ab} = |\varphi_{ab}|^2 \,, \quad P_{bc} = |\varphi_{bc}|^2 \,, \text{ and } P^q_{ac} = |\varphi_{ac}|^2 \,. \tag{3}$$

[5] We have assumed b is a non-degenerate state, and that therefore (1) is true. Presumably, if in some generalization of quantum mechanics (1) were not true, even for pure states b, (2) could be expected to be replaced by: There are complex numbers $P_{abc} = |\varphi_{abc}|^2$. The analog of (5) is then $\varphi_{ac} = \Sigma_b \, \varphi_{abc}$.

The classical law, obtained by combining (1) and (2),

$$P_{ac} = \sum_b P_{ab} P_{bc} \tag{4}$$

is replaced by

$$\varphi_{ac} = \sum_b \varphi_{ab} \varphi_{bc} \,. \tag{5}$$

If (5) is correct, ordinarily (4) is incorrect. The logical error made in deducing (4) consisted, of course, in assuming that to get from a to c the system had to go through a condition such that B had to have some definite value, b.

If an attempt is made to verify this, i.e., if B is measured between the experiments A and C, then formula (4) is, in fact, correct. More precisely, if the apparatus to measure B is set up and used, but no attempt is made to utilize the results of the B measurement in the sense that only the A to C correlation is recorded and studied, then (4) is correct. This is because the B measuring machine has done its job; if we wish, we could read the meters at any time without disturbing the situation any further. The experiments which gave a and c can, therefore, be separated into groups depending on the value of b.

Looking at probability from a frequency point of view (4) simply results from the statement that in each experiment giving a and c, B had some value. The only way (4) could be wrong is the statement, "B had some value," must sometimes be meaningless. Noting that (5) replaces (4) only under the circumstance that we make no attempt to measure B, we are led to say that the statement, "B had some value," may be meaningless whenever we make no attempt to measure B.[6]

Hence, we have different results for the correlation of a and c, namely, Eq. (4) or Eq. (5), depending upon whether we do or do not attempt to measure B. No matter how subtly one tries, the attempt to measure B must disturb the system, at least enough to change the results

[6] It does not help to point out that we *could* have measured B had we wished. The fact is that we did not.

from those given by (5) to those of (4).[7] That measurements do, in fact, cause the necessary disturbances, and that, essentially, (4) could be false was first clearly enunciated by Heisenberg in his uncertainty principle. The law (5) is a result of the work of Schroedinger, the statistical interpretation of Born and Jordan, and the transformation theory of Dirac.[8]

Equation (5) is a typical representation of the wave nature of matter. Here, the chance of finding a particle going from a to c through several different routes (values of b) may, if no attempt is made to determine the route, be represented as the square of a sum of several complex quantities — one for each available route. Probability can show the typical phenomena of interference, usually associated with waves, whose intensity is given by the square of the sum of contributions from different sources. The electron acts as a wave, (5), so to speak, as long as no attempt is made to verify that it is a particle; yet one can determine, if one wishes, by what route it travels just as though it were a particle; but when one does that, (4) applies and it does act like a particle.

These things are, of course, well known. They have already been explained many times.[9] However, it seems worthwhile to emphasize the fact that they are all simply direct consequences of Eq. (5), for it is essentially Eq. (5) that is fundamental in my formulation of quantum mechanics.

The generalization of Eqs. (4) and (5) to a large number of measurements, say A, B, C, $D, \ldots, K$, is, of course, that the probability of the sequence a, b, c, $d, \ldots, k$ is

$$P_{abcd\ldots k} = |\varphi_{abcd\ldots k}|^2 .$$

[7] How (4) actually results from (5) when measurements disturb the system has been studied particularly by J. von Neumann (*Mathematische Grundlagen der Quantenmechanik* (Dover Publications, New York, 1943)). The effect of perturbation of the measuring equipment is effectively to change the phase of the interfering components, by θ_b, say, so that (5) becomes $\psi_{ac} = \Sigma_b e^{i\theta_b}\varphi_{ab}\varphi_{bc}$. However, as von Neumann shows, the phase shifts must remain unknown if B is measured so that the resulting probability P_{ac} is the square of ψ_{ac} averaged over all phases, θ_b. This results in (4).

[8] If $\mathbf{A}$ and $\mathbf{B}$ are the operators corresponding to measurements A and B, and if ψ_a and ψ_b are solutions of $\mathbf{A}\psi_a = a\psi_a$ and $\mathbf{B}\chi_b = b\chi_b$, then $\varphi_{ab} = \int \chi_b^* \psi_a dx = (\chi_b^*, \psi_a)$. Thus, φ_{ab} is an element $(a|b)$ of the transformation matrix for the transformation from a representation in which $\mathbf{A}$ is diagonal to one in which $\mathbf{B}$ is diagonal.

[9] See, for example, W. Heisenberg, *The Physical Principles of the Quantum Theory* (University of Chicago Press, Chicago, 1930), particularly Chapter IV.

The probability of the result a, c, k, for example, if b, $d, \ldots$, are measured, is the classical formula:

$$P_{ack} = \sum_b \sum_d \cdots P_{abcd\ldots k} \,, \tag{6}$$

while the probability of the same sequence a, c, k if no measurements are made between A and C and between C and K is

$$P^q_{ack} = \left| \sum_b \sum_d \cdots \varphi_{abcd\cdots k} \right|^2 . \tag{7}$$

The quantity $\varphi_{abcd\cdots k}$ we can call the probability amplitude for the condition $A = a$, $B = b$, $C = c$, $D = d, \ldots, K = k$. (It is, of course, expressible as a product $\varphi_{ab}\varphi_{bc}\varphi_{cd} \cdots \varphi_{jk}$.)

3. The Probability Amplitude for a Space-Time Path

The physical ideas of the last section may be readily extended to define a probability amplitude for a particular completely specified space-time path. To explain how this may be done, we shall limit ourselves to a one-dimensional problem, as the generalization to several dimensions is obvious.

Assume that we have a particle which can take up various values of a coordinate x. Imagine that we make an enormous number of successive position measurements, let us say separated by a small time interval ε. Then a succession of measurements such as A, B, $C, \ldots$, might be the succession of measurements of the coordinate x at successive times $t_1, t_2, t_3, \ldots$, where $t_{i+1} = t_i + \varepsilon$. Let the value, which might result from measurement of the coordinate at time t_i, be x_i. Thus, if A is a measurement of x at t_1 then x_1 is what we previously denoted by a. From a classical point of view, the successive values, $x_1, x_2, x_3, \ldots$, of the coordinate practically define a path $x(t)$. Eventually, we expect to go the limit $\varepsilon \to 0$.

The probability of such a path is a function of $x_1, x_2, \ldots, x_i, \ldots$, say $P(\ldots x_i, x_{i+1}, \ldots)$. The probability that the path lies in a particular region R of space-time is obtained classically by integrating P over that region. Thus, the probability that x_i lies between a_i and b_i, and x_{i+1} lies between

a_{i+1} and b_{i+1}, etc., is

$$\cdots \int_{a_i}^{b_i} \int_{a_{i+1}}^{b_{i+1}} \cdots P(\ldots x_i, x_{i+1}, \ldots) \cdots dx_i dx_{i+1} \cdots$$

$$= \int_R P(\ldots x_i, x_{i+1}, \ldots) \cdots dx_i dx_{i+1} \cdots , \tag{8}$$

the symbol $\int_R$ meaning that the integration is to be taken over those ranges of the variables which lie within the region R. This is simply Eq. (6) with $a, b, \ldots,$ replaced by $x_1, x_2, \ldots,$ and integration replacing summation.

In quantum mechanics this is the correct formula for the case that $x_1, x_2, \ldots, x_i, \ldots,$ were actually all measured, and then only those paths lying within R were taken. We would expect the result to be different if no such detailed measurements had been performed. Suppose a measurement is made which is capable only of determining that the path lies somewhere within R.

The measurement is to be what we might call an "ideal measurement." We suppose that no further details could be obtained from the same measurement without further disturbance to the system. I have not been able to find a precise definition. We are trying to avoid the extra uncertainties that must be averaged over if, for example, more information were measured but not utilized. We wish to use Eq. (5) or Eq. (7) for all x_i and have no residual part to sum over in the manner of Eq. (4).

We expect that the probability that the particle is found by our "ideal measurement" to be, indeed, in the region R is the square of a complex number $|\varphi(R)|^2$. The number $\varphi(R)$, which we may call the probability amplitude for region R is given by Eq. (7) with $a, b, \ldots,$ replaced by $x_i, x_{i+1}, \ldots,$ and summation replaced by integration:

$$\varphi(R) = \operatorname*{Lim}_{\varepsilon \to 0} \int_R \Phi(\cdots x_i, x_{i+1} \cdots) \cdots dx_i dx_{i+1} \cdots . \tag{9}$$

The complex number $\Phi(\cdots x_i, x_{i+1} \cdots)$ is a function of the variables x_i defining the path. Actually, we imagine that the time spacing ε approaches zero so that Φ essentially depends on the entire path $x(t)$ rather than only on just the values of x_i at the particular times t_i, $x_i = x(t_i)$. We might call Φ the probability amplitude functional of paths $x(t)$.

We may summarize these ideas in our first postulate:

I. If an ideal measurement is performed to determine whether a particle has a path lying in a region of space-time, then the probability that the result will be affirmative is the absolute square of a sum of complex contributions, one from each path in the region.

The statement of the postulate is incomplete. The meaning of a sum of terms one for "each" path is ambiguous. The precise meaning given in Eq. (9) is this: A path is first defined only by the positions x_i through which it goes at a sequence of equally spaced times,[10] $t_i = t_{i-1} + \varepsilon$. Then all values of the coordinates within R have an equal weight. The actual magnitude of the weight depends upon ε and can be so chosen that the probability of an event which is certain shall be normalized to unity. It may not be best to do so, but we have left this weight factor in a proportionality constant in the second postulate. The limit $\varepsilon \to 0$ must be taken at the end of a calculation.

When the system has several degrees of freedom the coordinate space x has several dimensions so that the symbol x will represent a set of coordinates $(x^{(1)}, x^{(2)}, \ldots, x^{(k)})$ for a system with k degrees of freedom. A path is a sequence of configurations for successive times and is described by giving the configuration x_i or $(x_i^{(1)}, x_i^{(2)}, \ldots, x_i^{(k)})$, i.e., the value of each of the k coordinates for each time t_i. The symbol dx_i will be understood to mean the volume element in k-dimensional configuration space (at time t_i). The statement of the postulates is independent of the coordinate system which is used.

The postulate is limited to defining the results of position measurements. It does not say what must be done to define the result of a momentum measurement, for example. This is not a real limitation, however, because in principle the measurement of momentum of one particle can be performed in terms of position measurements of other particles, e.g., meter indicators. Thus, an analysis of such an experiment will determine what it is about the first particle which determines its momentum.

---

[10] There are very interesting mathematical problems involved in the attempt to avoid the subdivision and limiting processes. Some sort of complex measure is being associated with the space of functions $x(t)$. Finite results can be obtained under unexpected circumstances because the measure is not positive everywhere, but the contributions from most of the paths largely cancel out. These curious mathematical problems are sidestepped by the subdivision process. However, one feels as Cavalieri must have felt calculating the volume of a pyramid before the invention of calculus.

4. The Calculation of the Probability Amplitude for a Path

The first postulate prescribes the type of mathematical framework required by quantum mechanics for the calculation of probabilities. The second postulate gives a particular content to this framework by prescribing how to compute the important quantity Φ for each path:

II. The paths contribute equally in magnitude, but the phase of their contribution is the classical action (in units of $\hbar$); i.e., the time integral of the Lagrangian taken along the path.

That is to say, the contribution $\Phi[x(t)]$ from a given path $x(t)$ is proportional to $\exp(i/\hbar)S[x(t)]$, where the action $S[x(t)] = \int L(\dot{x}(t), x(t))dt$ is the time integral of the classical Lagrangian $L(\dot{x}, x)$ taken along the path is question. The Lagrangian, which may be an explicit function of the time, is a function of position and velocity. If we suppose it to be a quadratic function of the velocities, we can show the mathematical equivalence of the postulates here and the more usual formulation of quantum mechanics.

To interpret the first postulate it was necessary to define a path by giving only the succession of points x_i through which the path passes at successive times t_i. To compute $S = \int L(\dot{x}, x)dt$ we need to know the path at all points, not just at x_i. We shall assume that the function $x(t)$ in the interval between t_i and t_{i+1} is the path followed by a classical particle, with the Lagrangian L, which starting from x_i at t_i reaches x_{i+1} at t_{i+1}. This assumption is required to interpret the second postulate for discontinuous paths. The quantity $\Phi(\ldots x_i, x_{i+1}, \ldots)$ can be normalized (for various ε) if desired, so that the probability of an event which is certain is normalized to unity as $\varepsilon \to 0$.

There is no difficulty in carrying out the action integral because of the sudden changes of velocity encountered at the times t_i as long as L does not depend upon any higher time derivatives of the position than the first. Furthermore, unless L is restricted in this way the end points are not sufficient to define the classical path. Since the classical path is the one which makes the action a minimum, we can write

$$S = \sum_i S(x_{i+1}, x_i)\,, \tag{10}$$

where

$$S(x_{i+1}, x_i) = \text{Min} \int_{t_i}^{t_{i+1}} L(\dot{x}(t), x(t))dt \,. \tag{11}$$

Written in this way, the only appeal to classical mechanics is to supply us with a Lagrangian function. Indeed, one could consider postulate two as simply saying, "Φ is the exponential of i times the integral of a real function of $x(t)$ and its first time derivative." Then the classical equations of motion might be derived later as the limit for large dimensions. The function of x and $\dot{x}$ then could be shown to be the classical Lagrangian within a constant factor.

Actually, the sum in (10), even for finite ε, is infinite and hence meaningless (because of the infinite extent of time). This reflects a further incompleteness of the postulates. We shall have to restrict ourselves to a finite, but arbitrarily long, time interval.

Combining the two postulates and using Eq. (10), we find

$$\varphi(R) = \lim_{\varepsilon \to 0} \int_R \exp\left[\frac{i}{\hbar}\Sigma_i S(x_{i+1}, x_i)\right] \cdots \frac{dx_{i+1}}{A}\frac{dx_i}{A} \cdots \,, \tag{12}$$

where we have let the normalization factor be split into a factor $1/A$ (whose exact value we shall presently determine) for each instant of time. The integration is just over those values $x_i, x_{i+1}, \ldots$ which lie in the region R. This equation, the definition (11) of $S(x_{i+1}, x_i)$, and the physical interpretation of $|\varphi(R)|^2$ as the probability that the particle will be found in R, complete our formulation of quantum mechanics.

5. Definition of the Wave Function

We now proceed to show the equivalence of these postulates to the ordinary formulation of quantum mechanics. This we do in two steps. We show in this section how the wave function may be defined from the new point of view. In the next section we shall show that this function satisfies Schroedinger's differential wave equation.

We shall see that it is the possibility, (10), of expressing S as a sum, and hence Φ as a product, of contributions from successive sections of the path, which leads to the possibility of defining a quantity having the properties of a wave function.

To make this clear, let us imagine that we choose a particular time t and divide the region R in Eq. (12) into pieces, future and past relative to t. We imagine that R can be split into: (a) a region R', restricted in any way in space, but lying entirely earlier in time than some t', such that $t' < t$; (b) a region R'' arbitrarily restricted in space but lying entirely later in time than t'', such that $t'' > t$; (c) the region between t' and t'' in which all the values of x coordinates are unrestricted, i.e., all of space-time between t' and t''. The region (c) is not absolutely necessary. It can be taken as narrow in time as desired. However, it is convenient in letting us consider varying t a little without having to redefine R' and R''. Then $|\varphi(R', R'')|^2$ is the probability that the path occupies R' and R''. Because R' is entirely previous to R'', considering the time t as the present, we can express this as the probability that the path had been in region R' and will be in region R''. If we divide by a factor, the probability that the path is in R', to renormalize the probability we find: $|\varphi(R', R'')|^2$ is the (relative) probability that if the system were in region R' it will be found later in R''.

This is, of course, the important quantity in predicting the results of many experiments. We prepare the system in a certain way (e.g., it was in region R') and then measure some other property (e.g., will it be found in region R''?). What does (12) say about computing this quantity, or rather the quantity $\varphi(R', R'')$ of which it is the square?

Let us suppose in Eq. (12) that the time t corresponds to one particular point k of the subdivision of time into steps ε, i.e., assume $t = t_k$, the index k, of course, depending upon the subdivision ε. Then, the exponential being the exponential of a sum may be split into a product of two factors

$$\exp\left[\frac{i}{\hbar}\sum_{i=k}^{\infty} S(x_{i+1}, x_i)\right] \cdot \exp\left[\frac{i}{\hbar}\sum_{i=-\infty}^{k-1} S(x_{i+1}, x_i)\right] . \tag{13}$$

The first factor contains only coordinates with index k or higher, while the second contains only coordinates with index k or lower. This split is possible because of Eq. (10), which results essentially from the fact that the Lagrangian is a function only of positions and velocities. First, the integration on all variables x_i for $i > k$ can be performed on the first factor resulting in a function of x_k (times the second factor). Next, the integration on all variables x_i for $i < k$ can be performed on the second factor also, giving a function of x_k. Finally, the integration on x_k can be performed.

That is, $\varphi(R', R'')$ can be written as the integral over x_k of the product of two factors. We will call these $\chi^*(x_k, t)$ and $\psi(x_k, t)$:

$$\varphi(R', R'') = \int \chi^*(x, t)\psi(x, t)dx\,, \tag{14}$$

where

$$\psi(x_k, t) = \operatorname*{Lim}_{\varepsilon\to 0} \int_{R'} \exp\left[\frac{i}{\hbar}\sum_{i=-\infty}^{k-1} S(x_{i+1}, x_i)\right] \frac{dx_{k-1}}{A}\frac{dx_{k-2}}{A}\cdots\,, \tag{15}$$

and

$$\chi^*(x_k, t) = \operatorname*{Lim}_{\varepsilon\to 0} \int_{R''} \exp\left[\frac{i}{\hbar}\sum_{i=k}^{\infty} S(x_{i+1}, x_i)\right] \cdot \frac{1}{A}\frac{dx_{k+1}}{A}\frac{dx_{k+2}}{A}\cdots\,. \tag{16}$$

The symbol R' is placed on the integral for ψ to indicate that the coordinates are integrated over the region R', and, for t_i between t' and t, over all space. In like manner, the integral for χ^* is over R'' and over all space for those coordinates corresponding to times between t and t''. The asterisk on χ^* denotes complex conjugate, as it will be found more convenient to define (16) as the complex conjugate of some quantity, χ.

The quantity ψ depends only upon the region R' previous to t, and is completely defined if that region is known. It does not depend, in any way, upon what will be done to the system after time t. This latter information is contained in χ. Thus, with ψ and χ we have separated the past history from the future experiences of the system. This permits us to speak of the relation of past and future in the conventional manner. Thus, if a particle has been in a region of space-time R' it may at time t be said to be in a certain condition, or state, determined only by its past and described by the so-called were function $\psi(x, t)$. This function contains all that is needed to predict future probabilities. For suppose, in another situation, the region R' were different, say r', and possibly the Lagrangian for times before t were also altered. But, neverthelss, suppose the quantity from Eq. (15) turned out to be the same. Then, according to (14) the probability of ending in any region R'' is the same for R' as for r'. Therefore, future measurements will not distinguish whether the system had occupied R' or r'. Thus, the wave function $\psi(x, t)$ is sufficient to define those attributes which are left from past history which determine future behavior.

Likewise, the function $\chi^*(x,t)$ characterizes the experience, or, let us say, experiment to which the system is to be subjected. If a different region, r'' and different Lagrangian after t, were to give the same $\chi^*(x,t)$ *via* Eq. (16), as does region R'', then no matter what the preparation, ψ, Eq. (14) says that the chance of finding the system in R'' is always the same as finding it in r''. The two "experiments" R'' and r'' are equivalent, as they yield the same results. We shall say loosely that these experiments are to determine with what probability the system is in state χ. Actually, this terminology is poor. The system is really in state ψ. The reason we can associate a state with an experiment is, of course, that for an ideal experiment there turns out to be a unique state (whose wave function is $\chi(x,t)$) for which the experiment succeeds with certainty.

Thus, we can say: the probability that a system in state ψ will be found by an experiment whose characteristic state is χ (or, more loosely, the chance that a system in state ψ will appear to be in χ) is

$$\left| \int \chi^*(x,t)\psi(x,t)dx \right|^2 . \tag{17}$$

These results agree, of course, with the principles of ordinary quantum mechanics. They are a consequence of the fact that the Lagrangian is a function of position, velocity, and time only.

6. The Wave Equation

To complete the proof of the equivalence with the ordinary formulation we shall have to show that the wave function defined in the previous section by Eq. (15) actually satisfies the Schroedinger wave equation. Actually, we shall only succeed in doing this when the Lagrangian L in (11) is a quadratic, but perhaps inhomogeneous, form in the velocities $\dot{x}(t)$. This is not a limitation, however, as it includes all the cases for which the Schroedinger equation has been verified by experiment.

The wave equation describes the development of the wave function with time. We may expect to approach it by noting that, for finite ε, Eq. (15) permits a simple recursive relation to be developed. Consider the appearance of Eq. (15) if we were to compute ψ at the next instant of time:

$$\psi(x_{k+1}, t+\varepsilon) = \int_{R^1} \exp\left[\frac{i}{\hbar}\sum_{i=-\infty}^{k} S(x_{i+1}, x_i)\right] \frac{dx_k}{A}\frac{dx_{k-1}}{A}\cdots . \tag{15'}$$

This is similar to (15) except for the integration over the additional variable x_k and the extra term in the sum in the exponent. This term means that the integral of (15′) is the same as the integral of (15) except for the factor $(1/A)\exp(i/\hbar)S(x_{k+1}, x_k)$. Since this does not contain any of the variables x_i for i less than k, all of the integrations on dx_i up to dx_{k-1} can be performed with this factor left out. However, the result of these integrations is by (15) simply $\psi(x_k, t)$. Hence, we find from (15′) the relation

$$\psi(x_{k+1}, t+\varepsilon) = \int \exp\left[\frac{i}{\hbar}S(x_{k+1}, x_k)\right]\psi(x_k, t)dx_k/A\,. \tag{18}$$

This relation giving the development of ψ with time will be shown, for simple examples, with suitable choice of A, to be equivalent to Schroedinger's equation. Actually, Eq. (18) is not exact, but is only true in the limit $\varepsilon \to 0$ and we shall derive the Schroedinger equation by assuming (18) is valid for first order in ε. Equation (18) *need* only be true for small ε to the first order in ε. For if we consider the factors in (15) which carry us over a finite interval of time, T, the number of factors is T/ε. If an error of order ε^2 is made in each, the resulting error will not accumulate beyond the order $\varepsilon^2(T/\varepsilon)$ or T_ε, which vanishes in the limit.

We shall illustrate the relation of (18) to Schroedinger's equation by applying it to the simple case of a particle moving in one dimension in a potential $V(x)$. Before we do this, however, we would like to discuss some approximations to the value $S(x_{i+1}, x_i)$ given in (11) which will be sufficient for expression (18).

The expression defined in (11) for $S(x_{i+1}, x_i)$ is difficult to calculate exactly for arbitrary ε from classical mechanics. Actually, it is only necessary that an approximate expression for $S(x_{i+1}, x_i)$ be used in (18), provided the error of the approximation be of an order smaller than the first in ε. We limit ourselves to the case that the Lagrangian is a quadratic, but perhaps inhomogeneous, form in the velocities $\dot{x}(t)$. As we shall see later, the paths which are important are those for which $x_{i+1} - x_i$ is of order $\varepsilon^{\frac{1}{2}}$. Under these circumstances, it is sufficient to calculate the integral in (11) over the classical path taken by a *free* particle.[11]

In *Cartesian coordinates*[12] the path of a free particle is a straight line so the integral of (11) can be taken along a straight line. Under these circumstances it is sufficiently accurate to replace the integral by the trapezoidal rule

$$S(x_{i+1}, x_i) = \frac{\varepsilon}{2} L\left(\frac{x_{i+1} - x_i}{\varepsilon}, x_{i+1}\right) + \frac{\varepsilon}{2} L\left(\frac{x_{i+1} - x_i}{\varepsilon}, x_i\right) \tag{19}$$

or, if it proves more convenient,

$$S(x_{i+1}, x_i) = \varepsilon L\left(\frac{x_{i+1} - x_i}{\varepsilon}, \frac{x_{i+1} + x_i}{2}\right). \tag{20}$$

These are not valid in a general coordinate system, e.g., spherical. An even simpler approximation may be used if, in addition, there is no vector potential or other terms linear in the velocity (see p. 89):

$$S(x_{i+1}, x_i) = \varepsilon L\left(\frac{x_{i+1} - x_i}{\varepsilon}, x_{i+1}\right). \tag{21}$$

Thus, for the simple example of a particle of mass m moving in one dimension under a potential $V(x)$, we can set

$$S(x_{i+1}, x_i) = \frac{m\varepsilon}{2}\left(\frac{x_{i+1} - x_i}{\varepsilon}\right)^2 - \varepsilon V(x_{i+1}). \tag{22}$$

For this example, then, Eq. (18) becomes

$$\psi(x_{k+1}, t+\varepsilon) = \int \exp\left[\frac{i\varepsilon}{\hbar}\left\{\frac{m}{2}\left(\frac{x_{k+1} - x_k}{\varepsilon}\right)^2 - V(x_{k+1})\right\}\right]\psi(x_k, t)dx_k/A. \tag{23}$$

Let us call $x_{k+1} = x$ and $x_{k+1} - x_k = \xi$ so that $x_k = x - \xi$. Then (23)

[11] It is assumed that the "forces" enter through a scalar and vector potential and not in terms involving the square of the velocity. More generally, what is meant by a free particle is one for which the Lagrangian is altered by omission of the terms linear in, and those independent of, the velocities.

[12] More generally, coordinates for which the terms quadratic in the velocity in $L(\dot{x}, x)$ appear with constant coefficients.

becomes

$$\psi(x, t+\varepsilon) = \int \exp \frac{im\xi^2}{\varepsilon \cdot 2\hbar} \cdot \exp \frac{-i\varepsilon V(x)}{\hbar} \cdot \psi(x-\xi, t) \frac{d\xi}{A}. \tag{24}$$

The integral on ξ will converge if $\psi(x, t)$ falls off sufficiently for large x (certainly if $\int \psi^*(x)\psi(x)dx = 1$). In the integration of ξ, since ε is very small, the exponential of $im\xi^2/2\hbar\varepsilon$ oscillates extremely rapidly except in the region about $\xi = 0$ (ξ of order $(\hbar\varepsilon/m)^{\frac{1}{2}}$). Since the function $\psi(x-\xi, t)$ is a relatively smooth function of ξ (since ε may be taken as small as desired), the region where the exponential oscillates rapidly will contribute very little because of the almost complete cancelation of positive and negative contributions. Since only small ξ are effective, $\psi(x-\xi, t)$ may be expanded as a Taylor series. Hence,

$$\psi(x, t+\varepsilon) = \exp\left(\frac{-i\varepsilon V(x)}{\hbar}\right) \int \exp\left(\frac{im\xi^2}{2\hbar\varepsilon}\right) \times \left[\psi(x,t) - \xi\frac{\partial\psi(x,t)}{\partial x} + \frac{\xi^2}{2}\frac{\partial^2\psi(x,t)}{\partial x^2} - \cdots\right] d\xi/A. \tag{25}$$

Now

$$\begin{aligned}
\int_{-\infty}^{\infty} \exp(im\xi^2/2\hbar\varepsilon)d\xi &= (2\pi\hbar\varepsilon i/m)^{\frac{1}{2}}, \\
\int_{-\infty}^{\infty} \exp(im\xi^2/2\hbar\varepsilon)\xi d\xi &= 0, \\
\int_{-\infty}^{\infty} \exp(im\xi^2/2\hbar\varepsilon)\xi^2 d\xi &= (\hbar\varepsilon i/m)(2\pi\hbar\varepsilon i/m)^{\frac{1}{2}},
\end{aligned} \tag{26}$$

while the integral containing ξ^3 is zero, for like the one with ξ it possesses an odd integrand, and the ones with ξ^4 are of at least the order ε smaller than the ones kept here.[13] If we expand the left-hand side to first order

---

[13] Really, these integrals are oscillatory and not defined, but they may be defined by using a convergence factor. Such a factor is automatically provided by $\psi(x-\varepsilon, t)$ in (24). If a more formal procedure is desired replace $\hbar$ by $\hbar(1-i\delta)$, for example, where δ is a small positive number, and then let $\delta \to 0$.

in ε, (25) becomes

$$\psi(x,t) + \varepsilon\frac{\partial\psi(x,t)}{\partial t} = \exp\left(\frac{-i\varepsilon V(x)}{\hbar}\right)\frac{(2\pi\hbar\varepsilon i/m)^{\frac{1}{2}}}{A} \times \left[\psi(x,t) + \frac{\hbar\varepsilon i}{m}\frac{\partial^2\psi(x,t)}{\partial x^2} + \cdots\right]. \tag{27}$$

In order that both sides may agree to *zero* order in ε, we must set

$$A = (2\pi\hbar\varepsilon i/m)^{\frac{1}{2}}. \tag{28}$$

Then expanding the exponential containing $V(x)$, we get

$$\psi(x,t) + \varepsilon\frac{\partial\psi}{\partial t} = \left(1 - \frac{i\varepsilon}{\hbar}V(x)\right)\left(\psi(x,t) + \frac{\hbar\varepsilon i}{2m}\frac{\partial^2\psi}{\partial x^2}\right). \tag{29}$$

Canceling $\psi(x,t)$ from both sides, and comparing terms to first order in ε and multiplying by $-\hbar/i$ one obtains

$$-\frac{\hbar}{i}\frac{\partial\psi}{\partial t} = \frac{1}{2m}\left(\frac{\hbar}{i}\frac{\partial}{\partial x}\right)^2\psi + V(x)\psi, \tag{30}$$

which is Schroedinger's equation for the problem in question.

The equation for χ^* can be developed in the same way, but adding a factor *decreases* the time by one step, i.e., χ^* satisfies an equation like (30) but with the sign of the time reversed. By taking complex conjugates we can conclude that χ satisfies the same equation as ψ, i.e., an experiment can be defined by the particular state χ to which it corresponds.[14]

This example shows that most of the contribution to $\psi(x_{k+1}, t+\varepsilon)$ comes from values of x_k in $\psi(x_k, t)$ which are quite close to x_{k+1} (distant of order $\varepsilon^{\frac{1}{2}}$) so that the integral equation (23) can, in the limit, be replaced by a differential equation. The "velocities," $(x_{k+1} - x_k)/\varepsilon$ which are important are very high, being of order $(\hbar/m\varepsilon)^{\frac{1}{2}}$ which diverges as $\varepsilon \to 0$. The paths involved are, therefore continuous but possess no derivative. They are of a type familiar from study of Brownian motion.

[14] Dr. Hartland Snyder has pointed out to me, in private conversation, the very interesting possibility that there may be a generalization of quantum mechanics in which the states measured by experiment cannot be prepared; that is, there would be no state into which a system may be put for which a particular experiment gives certainty for a result. The class of functions χ is not identical to the class of available states ψ. This would result if, for example, χ satisfied a different equation than ψ.

It is these large velocities which make it so necessary to be careful in approximating $S(x_{k+1}, x_k)$ from Eq. (11).[15] To replace $V(x_{k+1})$ by $V(x_k)$ would, of course, change the exponent in (18) by $i\varepsilon[V(x_k) - V(x_{k+1})]/\hbar$ which is of order $\varepsilon(x_{k+1} - x_k)$, and thus lead to unimportant terms of higher order than ε on the right-hand side of (29). It is for this reason that (20) and (21) are equally satisfactory approximations to $S(x_{i+1}, x_i)$ when there is no vector potential. A term, linear in velocity, however, arising from a vector potential, as $A\dot{x}\,dt$ must be handled more carefully. Here a term in $S(x_{k+1}, x_k)$ such as $A(x_{k+1})(x_{k+1} - x_k)$ differs from $A(x_k)(x_{k+1} - x_k)$ by a term of order $(x_{k+1} - x_k)^2$, and, therefore, of order ε. Such a term would lead to a change in the resulting wave equation. For this reason the approximation (21) is not a sufficiently accurate approximation to (11) and one like (20), (or (19) from which (20) differs by terms of order higher than ε) must be used. If $\mathbf{A}$ represents the vector potential and $\mathbf{p} = (\hbar/i)\nabla$, the momentum operator, then (20) gives, in the Hamiltonian operator, a term $(1/2m)(\mathbf{p} - (e/c)\mathbf{A}) \cdot (\mathbf{p} - (e/c)\mathbf{A})$, while (21) gives $(1/2m)(\mathbf{p} \cdot \mathbf{p} - (2e/c)\mathbf{A} \cdot \mathbf{p} + (e^2/c^2)\mathbf{A} \cdot \mathbf{A})$. These two expressions differ by $(\hbar e/2imc)\nabla \cdot \mathbf{A}$ which may not be zero. The question is still more important in the coefficient of terms which are quadratic in the velocities. In these terms (19) and (20) are not sufficiently accurate representations of (11) in general. It is when the coefficients are constant that (19) or (20) can be substituted for (11). If an expression such as (19) is used, say for spherical coordinates, when it is not a valid approximation to (11), one obtains a Schroedinger equation in which the Hamiltonian operator has some of the momentum operators and coordinates in the wrong order. Equation (11) then resolves the ambiguity in the usual rule to replace p and q by the non-commuting quantities $(\hbar/i)(\partial/\partial q)$ and q in the classical Hamiltonian $H(p, q)$.

It is clear that the statement (11) is independent of the coordinate system. Therefore, to find the differential wave equation it gives in any

---

[15] Equation (18) is actually exact when (11) is used for $S(x_{i+1}, x_i)$ for arbitrary ε for cases in which the potential does not involve x to higher powers than the second (e.g., free particle, harmonic oscillator). It is necessary, however, to use a more accurate value of A. One can define A in this way. Assume classical particles with k degrees of freedom start from the point x_i, t_i with uniform density in momentum space. Write the number of particles having a given component of momentum in range dp as dp/p_0 with p_0 constant. Then $A = (2\pi\hbar i/p_0)^{k/2}\rho^{-\frac{1}{2}}$, where ρ is the density in k-dimensional coordinate space x_{i+1} of these particles at time t_{i+1}.

coordinate system, the easiest procedure is first to find the equations in Cartesian coordinates and then to transform the coordinate system to the one desired. It suffices, therefore, to show the relation of the postulates and Schroedinger's equation in rectangular coordinates.

The derivation given here for one dimension can be extended directly to the case of three-dimensional Cartesian coordinates for any number, K, of particles interacting through potentials with one another, and in a magnetic field, described by a vector potential. The terms in the vector potential require completing the square in the exponent in the usual way for Gaussian integrals. The variable x must be replaced by the set $x^{(1)}$ to $x^{(3K)}$ where $x^{(1)}$, $x^{(2)}$, $x^{(3)}$ are the coordinates of the first particle of mass m_1, $x^{(4)}$, $x^{(5)}$, $x^{(6)}$ of the second of mass m_2, etc. The symbol dx is replaced by $dx^{(1)}dx^{(2)}\cdots dx^{(3K)}$, and the integration over dx is replaced by a $3K$-fold integral. The constant A has, in this case, the value $A = (2\pi\hbar\varepsilon i/m_1)^{\frac{3}{2}}(2\pi\hbar\varepsilon i/m_2)^{\frac{3}{2}}\cdots(2\pi\hbar\varepsilon i/m_K)^{\frac{3}{2}}$. The Lagrangian is the classical Lagrangian for the same problem, and the Schroedinger equation resulting will be that which corresponds to the classical Hamiltonian, derived from this Lagrangian. The equations in any other coordinate system may be obtained by transformation. Since this includes all cases for which Schroedinger's equation has been checked with experiment, we may say our postulates are able to describe what can be described by non-relativistic quantum mechanics, neglecting spin.

7. Discussion of the Wave Equation: The Classical Limit

This completes the demonstration of the equivalence of the new and old formulations. We should like to include in this section a few remarks about the important equation (18).

This equation gives the development of the wave function during a small time interval. It is easily interpreted physically as the expression of Huygens' principle for matter waves. In geometrical optics the rays in an inhomogeneous medium satisfy Fermat's principle of least *time*. We may state Huygens' principle in wave optics in this way: If the amplitude of the wave is known on a given surface, the amplitude at a near by point can be considered as a sum of contributions from all points of the surface. Each contribution is delayed in phase by an amount proportional to the *time* it would take the light to get from the surface to the point along the ray of least *time* of geometrical optics. We can consider (22) in an analogous manner starting with Hamilton's first principle of least *action* for classical or "geometrical" mechanics. If the amplitude of the wave ψ is known on a given "surface," in particular the "surface" consisting of all x at time t, its value at a particular nearby point at time $t + \varepsilon$, is a sum of contributions from all points of the surface at t. Each contribution is delayed in phase by an amount proportional to the *action* it would require to get from the surface to the point along the path of least *action* of classical mechanics.[16]

Actually Huygens' principle is not correct in optics. It is replaced by Kirchoff's modification which requires that both the amplitude and its derivative must be known on the adjacent surface. This is a consequence of the fact that the wave equation in optics is second order in the time. The wave equation of quantum mechanics is first order in the time; therefore, Huygens' principle *is* correct for matter waves, action replacing time.

The equation can also be compared mathematically to quantities appearing in the usual formulations. In Schroedinger's method the development of the wave function with time is given by

$$-\frac{\hbar}{i}\frac{\partial\psi}{\partial t} = H\psi\,, \tag{31}$$

which has the solution (for any ε if $\mathbf{H}$ is time independent)

$$\psi(x, t + \varepsilon) = \exp(-i\varepsilon\mathbf{H}/\hbar)\psi(x, t)\,. \tag{32}$$

Therefore, Eq. (18) expresses the operator $\exp(-i\varepsilon\mathbf{H}\hbar)$ by an approximate integral operator for small ε.

From the point of view of Heisenberg one considers the position at time t, for example, as an operator $\mathbf{x}$. The position $\mathbf{x}'$ at a later time $t + \varepsilon$ can be expressed in terms of that at time t by the operator equation

$$\mathbf{x}' = \exp(i\varepsilon\mathbf{H}/\hbar)\mathbf{x}\exp-(i\varepsilon\mathbf{H}/\hbar)\,. \tag{33}$$

The transformation theory of Dirac allows us to consider the wave function at time $t + \varepsilon$, $\psi(x', t + \varepsilon)$, as representing a state in a representation in

[16] See in this connection the very interesting remarks of Schroedinger, *Ann. d. Physik* **79**, 489 (1926).

which $\mathbf{x}'$ is diagonal, while $\psi(x,t)$ represents the same state in a representation in which $\mathbf{x}$ is diagonal. They are, therefore, related through the transformation function $(x'|x)_\varepsilon$ which relates these representations:

$$\psi(x', t+\varepsilon) = \int (x'|x)_\varepsilon \psi(x,t) dx\,.$$

Therefore, the content of Eq. (18) is to show that for small ε we can set

$$(x'|x)_\varepsilon = (1/A)\exp(iS(x',x)/\hbar) \tag{34}$$

with $S(x',x)$ defined as in (11).

The close analogy between $(x'|x)_\varepsilon$ and the quantity $\exp(iS(x',x)/\hbar)$ has been pointed out on several occasions by Dirac.[1] In fact, we now see that to sufficient approximations the two quantities may be taken to be proportional to each other. Dirac's remarks were the starting point of the present development. The points he makes concerning the passage to the classical limit $\hbar \to 0$ are very beautiful, and I may perhaps the excused for briefly reviewing them here.

First we note that the wave function at x'' at time t'' can be obtained from that at x' at time t' by

$$\psi(x'',t'') = \operatorname*{Lim}_{\varepsilon\to 0} \int\cdots\int \exp\left[\frac{i}{\hbar}\sum_{i=0}^{j-1} S(x_{i+1},x_i)\right]\psi(x',t')\frac{dx_0}{A}\frac{dx_1}{A}\cdots\frac{dx_{j-1}}{A}\,, \tag{35}$$

where we put $x_0 \equiv x'$ and $x_j \equiv x''$ where $j\varepsilon = t'' - t'$ (between the times t' and t'' we assume no restriction is being put on the region of integration). This can be seen either by repeated applications of (18) or directly from Eq. (15). Now we ask, as $h \to 0$ what values of the intermediate coordinates x_i contribute most strongly to the integral? These will be the values most likely to be found by experiment and therefore will determine, in the limit, the classical path. If $\hbar$ is very small, the exponent will be a very rapidly varying function of any of its variables x_i. As x_i varies, the positive and negative contributions of the exponent nearly cancel. The region at which x_i contributes most strongly is that at which the phase of the exponent varies least rapidly with x_i (method of stationary phase). Call the sum in the exponent S;

$$S = \sum_{i=0}^{j-1} S(x_{i+1}, x_i) \,. \tag{36}$$

Then the classical orbit passes, approximately, through those points x_i at which the rate of change of S with x_i is small, or in the limit of small $\hbar$, zero, i.e., the classical orbit passes through the points at which $\partial S/\partial x_i = 0$ for all x_i. Taking the limit $\varepsilon \to 0$, (36) becomes in view of (11)

$$S = \int_{t'}^{t''} L(\dot{x}(t), x(t)) dt \,. \tag{37}$$

We see then that the classical path is that for which the integral (37) suffers no first-order change on varying the path. This is Hamilton's principle and leads directly to the Lagrangian equations of motion.

8. Operator Algebra: Matrix Elements

Given the wave function and Schroedinger's equation, of course all of the machinery of operator or matrix algebra can be developed. It is, however, rather interesting to express these concepts in a somewhat different language more closely related to that used in stating the postulates. Little will be gained by this in elucidating operator algebra. In fact, the results are simply a translation of simple operator equations into a somewhat more cumbersome notation. On the other hand, the new notation and point of view are very useful in certain applications described in the introduction. Furthermore, the form of the equations permits natural extension to a wider class of operators than is usually considered (e.g., ones involving quantities referring to two or more different times). If any generalization to a wider class of action functionals is possible, the formulae to be developed will play an important role.

We discuss these points in the next three sections. This section is concerned mainly with definitions. We shall define a quantity which we call a transition element between two states. It is essentially a matrix element. But instead of being the matrix element between a state ψ and another χ corresponding to the *same* time, these two states will refer to different times. In the following section a fundamental relation between transition elements will be developed from which the usual commutation rules between coordinate and momentum may be deduced. The same relation also yields Newton's equation of motion in matrix form. Finally, in Sec. 10 we discuss the relation of the Hamiltonian to the operation of displacement in time.

We begin by defining a transition element in terms of the probability of transition from one state to another. More precisely, suppose we have a situation similar to that described in deriving (17). The region R consists of a region R' previous to t', all space between t' and t'' and the region R'' after t''. We shall study the probability that a system in region R' is later found in region R''. This is given by (17). We shall discuss in this section how it changes with changes in the form of the Lagrangian between t' and t''. In Sec. 10 we discuss how it changes with changes in the preparation R' or the experiment R''.

The state at time t' is defined completely by the preparation R'. It can be specified by a wave function $\psi(x', t')$ obtained as in (15), but containing only integrals up to the time t'. Likewise, the state characteristic of the experiment (region R'') can be defined by a function $\chi(x'', t'')$ obtained from (16) with integrals only beyond t''. The wave function $\psi(x'', t'')$ at time t'' can, of course, also be gotten by appropriate use of (15). It can also be gotten from $\psi(x', t')$ by (35). According to (17) with t'' used instead of t, the probability of being found in χ if prepared in ψ is the square of what we shall call the transition amplitude $\int \chi^*(x'', t'')\psi(x'', t'')dx''$. We wish to express this is terms of χ at t'' and ψ at t'. This we can do with the aid of (35). Thus, the chance that a system prepared in state $\psi_{t'}$ at time t' will be found after t'' to be in a state $\chi_{t''}$ is the square of the transition amplitude

$$\langle \chi_{t''} | 1 | \psi_{t'} \rangle_S = \lim_{\varepsilon \to 0} \int \cdots \int \chi^*(x'', t'') \exp(iS/\hbar) \psi(x', t') \frac{dx_0}{A} \cdots \frac{dx_{j-1}}{A} dx_j \,, \tag{38}$$

where we have used the abbreviation (36).

In the language of ordinary quantum mechanics if the Hamiltonian, $\mathbf{H}$, is constant, $\psi(x, t'') = \exp[-i(t''-t')\mathbf{H}/\hbar]\psi(x, t')$ so that (38) is the matrix element of $\exp[-i(t''-t')\mathbf{H}/\hbar]$ between states $\chi_{t''}$ and $\psi_{t'}$.

If F is any function of the coordinates x_i for $t' < t_i < t''$, we shall define the transition element of F between the states ψ at t' and χ at t'' for the action S as $(x'' \equiv x_j, x' \equiv x_0)$:

$$\langle \chi_{t''}|F|\psi_{t'}\rangle_S = \lim_{\varepsilon\to 0}\int\cdots\int \chi^*(x'',t'')F(x_0,x_1,\ldots,x_j) \cdot \exp\left[\frac{i}{\hbar}\sum_{i=0}^{j-1} S(x_{i+1},x_i)\right]\psi(x',t')\frac{dx_0}{A}\cdots\frac{dx_{j-1}}{A}dx_j\,. \tag{39}$$

In the limit $\varepsilon \to 0$, F is a functional of the path $x(t)$.

We shall see presently why such quantities are important. It will be easier to understand if we stop for a moment to find out what the quantities correspond to in conventional notation. Suppose F is simply x_k where k corresponds to some time $t = t_k$. Then on the right-hand side of (39) the integrals from x_0 to x_{k-1} may be performed to produce $\psi(x_k,t)$ or $\exp[-i(t-t')\mathbf{H}/\hbar]\psi_{t'}$. In like manner the integrals on x_i for $j \geq i > k$ give $\chi^*(x_k,t)$ or $\{\exp[-i(t''-t)\mathbf{H}/\hbar]\chi_{t''}\}^*$. Thus, the transition element of x_k,

$$\begin{aligned}\langle \chi_{t''}|F|\psi_{t'}\rangle_S &= \int \chi^*_{t''}e^{-(i/\hbar)\mathbf{H}(t''-t)}xe^{-(i/\hbar)\mathbf{H}(t-t')}\psi_{t'}dx \\ &= \int \chi^*(x,t)x\psi(x,t)dx \end{aligned} \tag{40}$$

is the matrix element of $\mathbf{x}$ at time $t = t_k$ between the state which would develop at time t from $\psi_{t'}$ at t' and the state which will develop from time t to $\chi_{t''}$ at t''. It is, therefore, the matrix element of $\mathbf{x}(t)$ between these states.

Likewise, according to (39) with $F = x_{k+1}$, the transition element of x_{k+1} is the matrix element of $\mathbf{x}(t+\varepsilon)$. The transition element of $F = (x_{k+1}-x_k)/\varepsilon$ is the matrix element of $(\mathbf{x}(t+\varepsilon)-\mathbf{x}(t))/\varepsilon$ or of $i(\mathbf{Hx}-\mathbf{xH})/\hbar$, as is easily shown from (40). We can call this the matrix element of velocity $\dot{x}(t)$.

Suppose we consider a second problem which differs from the first because, for example, the potential is augmented by a small amount $U(,\mathbf{x}t)$. Then in the new problem the quantity replacing S is $S' = S + \Sigma_i \varepsilon U(x_i,t_i)$. Substitution into (38) leads directly to

$$\langle \chi_{t''}|1|\psi_{t'}\rangle_{S'} = \left\langle \chi_{t''}\left|\exp\frac{i\varepsilon}{\hbar}\sum_{i=1}^{j}U(x_i,t_i)\right|\psi_{t'}\right\rangle_S\,. \tag{41}$$

Thus, transition elements such as (39) are important insofar as F may arise in some way from a change δS in an action expression. We denote, by observable functionals, those functionals F which can be defined, (possibly indirectly) in terms of the changes which are produced by possible changes in the action S. The condition that a functional be observable is somewhat similar to the condition that an operator be Hermitian. The observable functionals are a restricted class because the action must remain a quadratic function of velocities. From one observable functional others may be derived, for example, by

$$\langle \chi_{t''}|F|\psi_{t'}\rangle_{S'} = \left\langle \chi_{t''} \left| F \exp \frac{i\varepsilon}{\hbar} \sum_{i=1}^{j} U(x_i, t_i) \right| \psi_{t'} \right\rangle_S \tag{42}$$

which is obtained from (39).

Incidentally, (41) leads directly to an important perturbation formula. If the effect of U is small the exponential can be expanded to first order in U and we find

$$\langle \chi_{t''}|1|\psi_{t'}\rangle_{S'} = \langle \chi_{t''}|1|\psi_{t'}\rangle_S + \frac{i}{\hbar} \left\langle \chi_{t''} \left| \sum_i \varepsilon U(x_i, t_i) \right| \psi_{t'} \right\rangle . \tag{43}$$

Of particular importance is the case that $\chi_{t''}$ is a state in which $\psi_{t'}$ would not be found at all were it not for the disturbance, U (i.e., $\langle \chi_{t''}|1|\psi_{t'}\rangle_S = 0$). Then

$$\frac{1}{\hbar^2} \left| \left\langle \chi_{t''} \left| \sum_i \varepsilon U(x_i, t_i) \right| \psi_{t'} \right\rangle_S \right|^2 \tag{44}$$

is the probability of transition as induced to first order by the perturbation. In ordinary notation,

$$\left\langle \chi_{t''} \left| \sum_i \varepsilon U(x_i, t_i) \right| \psi_{t'} \right\rangle_S = \int \left\{ \int \chi^*_{t''} e^{-(i/\hbar)\mathbf{H}(t''-t)} \mathbf{U} e^{-(i/\hbar)\mathbf{H}(t-t')} \psi_{t'} \, dx \right\} dt$$

so that (44) reduces to the usual expression[17] for time dependent perturbations.

[17] P. A. M. Dirac, *The Principles of Quantum Mechanics* (The Clarendon Press, Oxford, 1935), second edition, Sec. 47, Eq. (20).

9. Newton's Equations: The Commutation Relation

In this section we find that different functionals may give identical results when taken between any two states. This equivalence between functionals is the statement of operator equations in the new language.

If F depends on the various coordinates, we can, of course, define a new functional $\partial F/\partial x_k$ by differentiating it with respect to one of its variables, say $x_k (0 < k < j)$. If we calculate $\langle \chi_{t''} | \partial F/\partial x_k | \psi_{t'} \rangle_S$ by (39) the integral on the right-hand side will contain $\partial F/\partial x_k$. The only other place that the variable x_k appears is in S. Thus, the integration on x_k can be performed by parts. The integrated part vanishes (assuming wave functions vanish at infinity) and we are left with the quantity $-F(\partial/\partial x_k)\exp(iS/\hbar)$ in the integral. However, $(\partial/\partial x_k)\exp(iS/\hbar) = (i/\hbar)(\partial S/\partial x_k)\exp(iS/\hbar)$, so the right side represents the transition element of $-(i/\hbar)F(\partial S/\partial x_k)$, i.e.,

$$\left\langle \chi_{t''} \left| \frac{\partial F}{\partial x_k} \right| \psi_{t'} \right\rangle_S = -\frac{i}{\hbar} \left\langle \chi_{t''} \left| F \frac{\partial S}{\partial x_k} \right| \psi_{t'} \right\rangle_S . \tag{45}$$

This very important relation shows that two different functionals may give the same result for the transition element between any two states. We say they are equivalent and symbolize the relation by

$$-\frac{\hbar}{i} \frac{\partial F}{\partial x_k} \underset{S}{\leftrightarrow} F \frac{\partial S}{\partial x_k} , \tag{46}$$

the symbol $\underset{S}{\leftrightarrow}$ emphasizing the fact that functionals equivalent under one action may not be equivalent under another. The quantities in (46) need not be observable. The equivalence is, nevertheless, true. Making use of (36) one can write

$$-\frac{\hbar}{i} \frac{\partial F}{\partial x_k} \underset{S}{\leftrightarrow} F \left[\frac{\partial S(x_{k+1}, x_k)}{\partial x_k} + \frac{\partial S(x_k, x_{k+1})}{\partial x_k} \right] . \tag{47}$$

This equation is true to zero and first order in ε and has as consequences the commutation relations of momentum and coordinate, as well as the Newtonian equations of motion in matrix form.

In the case of our simple one-dimensional problem, $S(x_{i+1}, x_i)$ is given by the expression (15), so that

$$\partial S(x_{k+1}, x_k)/\partial x_k = -m(x_{k+1} - x_k)/\varepsilon ,$$

and

$$\partial S(x_k, x_{k+1})/\partial x_k = +m(x_k - x_{k-1})/\varepsilon - \varepsilon V'(x_k)\,;$$

where we write $V'(x)$ for the derivative of the potential, or force. Then (47) becomes

$$-\frac{\hbar}{i}\frac{\partial F}{\partial x_k} \underset{S}{\leftrightarrow} F\left[-m\left(\frac{x_{k+1}-x_k}{\varepsilon} - \frac{x_k - x_{k-1}}{\varepsilon}\right) - \varepsilon V'(x_k)\right]. \qquad (48)$$

If F does not depend on the variable x_k, this gives Newton's equations of motion. For example, if F is constant, say unity, (48) just gives (dividing by ε)

$$0 \underset{S}{\leftrightarrow} -\frac{m}{\varepsilon}\left(\frac{x_{k+1}-x_k}{\varepsilon} - \frac{x_k - x_{k-1}}{\varepsilon}\right) - V'(x_k)\,.$$

Thus, the transition element of mass times acceleration $[(x_{k+1} - x_k)/\varepsilon - (x_k - x_{k-1})/\varepsilon]/\varepsilon$ between any two states is equal to the transition element of force $-V'(x_k)$ between the same states. This is the matrix expression of Newton's law which holds in quantum mechanics.

What happens if F does depend upon x_k? For example, let $F = x_k$. Then (48) gives, since $\partial F/\partial x_k = 1$,

$$-\frac{\hbar}{i} \underset{S}{\leftrightarrow} x_k\left[-m\left(\frac{x_{k+1}-x_k}{\varepsilon} - \frac{x_k - x_{k-1}}{\varepsilon}\right) - \varepsilon V'(x_k)\right]$$

or, neglecting terms of order ε,

$$m\left(\frac{x_{k+1}-x_k}{\varepsilon}\right)x_k - m\left(\frac{x_k - x_{k-1}}{\varepsilon}\right)x_k \underset{S}{\leftrightarrow} \frac{\hbar}{i}\,. \qquad (49)$$

In order to transfer an equation such as (49) into conventional notation, we shall have to discover what matrix corresponds to a quantity such as $x_k x_{k+1}$. It is clear from a study of (39) that if F is set equal to, say, $f(x_k)g(x_{k+1})$, the corresponding operator in (40) is

$$e^{-(i/\hbar)(t''-t-\varepsilon)\mathbf{H}} g(\mathbf{x}) e^{-(i/\hbar)\varepsilon\mathbf{H}} f(\mathbf{x}) e^{-(i/\hbar)(t-t')\mathbf{H}}\,,$$

the matrix element being taken between the states $\chi_{t''}$ and $\psi_{t'}$. The operators corresponding to functions of x_{k+1} will appear to the left of the operators corresponding to functions of x_k, i.e., *the order of terms in a matrix operator product corresponds to an order in time of the corresponding factors in a functional.* Thus, if the functional can and is written in such a way that in each term factors corresponding to later times appear to the left of factors corresponding to earlier terms, the corresponding operator can immediately be written down if the order of the operators is kept the same as in the functional.[18] Obviously, the order of factors in a functional is of no consequence. The ordering just facilitates translation into conventional operator notation. To write Eq. (49) in the way desired for easy translation would require the factors in the second term on the left to be reversed in order. We see, therefore, that it corresponds to

$$\mathbf{p}\mathbf{x} - \mathbf{x}\mathbf{p} = \hbar/i$$

where we have written $\mathbf{p}$ for the operator $m\dot{\mathbf{x}}$.

The relation between functionals and the corresponding operators is defined above in terms of the order of the factors in time. It should be remarked that this rule must be especially carefully adhered to when quantities involving velocities or higher derivatives are involved. The correct functional to represent the operator $(\dot{x})^2$ is actually $(x_{k+1}-x_k)/\varepsilon\cdot(x_k-x_{k-1})/\varepsilon$ rather than $[(x_{k+1} - x_k)/\varepsilon]^2$. The latter quantity diverges as $1/\varepsilon$ as $\varepsilon \to 0$. This may be seen by replacing the second term in (49) by its value $x_{k+1} \cdot m(x_{k+1} - x_k)/\varepsilon$ calculated an instant ε later in time. This does not change the equation to zero order in ε. We then obtain (dividing by ε)

$$\left(\frac{x_{k+1} - x_k}{\varepsilon}\right)^2 \underset{S}{\leftrightarrow} -\frac{\hbar}{im_\varepsilon}. \tag{50}$$

This gives the result expressed earlier that the root mean square of the "velocity" $(x_{k+1} - x_k)/\varepsilon$ between two successive positions of the path is of order $\varepsilon^{-\frac{1}{2}}$.

It will not do then to write the functional for kinetic energy, say, simply as

$$\frac{1}{2}m[(x_{k+1} - x_k)/\varepsilon]^2 \tag{51}$$

for this quantity is infinite as $\varepsilon \to 0$. In fact, it is not an observable functional.

---

[18] Dirac has also studied operators containing quantities referring to different times. See reference 2.

One can obtain the kinetic energy as an observable functional by considering the first-order change in transition amplitude occasioned by a change in the mass of the particle. Let m be changed to $m(1+\delta)$ for a short time, say ε, around t_k. The change in the action is $\frac{1}{2}\delta\varepsilon m[(x_{k+1}-x_k)/\varepsilon]^2$ the derivative of which gives an expression like (51). But the change in m changes the normalization constant $1/A$ corresponding to dx_k as well as the action. The constant is changed from $(2\pi\hbar\varepsilon i/m)^{-\frac{1}{2}}$ to $(2\pi\hbar\varepsilon i/m(1+\delta))^{-\frac{1}{2}}$ or by $\frac{1}{2}\delta(2\pi\hbar\varepsilon i/m)^{-\frac{1}{2}}$ to first order in δ. The total effect of the change in mass in Eq. (38) to the first order in δ is

$$\left\langle \chi_{t''} \left| \frac{1}{2}\delta\varepsilon i m[(x_{k+1}-x_k)/\varepsilon]^2/\hbar + \frac{1}{2}\delta \right| \psi_{t'} \right\rangle .$$

We expect the change of order δ lasting for a time ε to be of order $\delta\varepsilon$. Hence, dividing by $\delta\varepsilon i/\hbar$, we can define the kinetic energy functional as

$$\text{K.E.} = \frac{1}{2}m[(x_{k+1}-x_k)/\varepsilon]^2 + \hbar/2\varepsilon i\,. \tag{52}$$

This is finite as $\varepsilon \to 0$ in view of (50). By making use of an equation which results from substituting $m(x_{k+1}-x_k)/\varepsilon$ for F in (48) we can also show that the expression (52) is equal (to order ε) to

$$\text{K.E.} = \frac{1}{2}m\left(\frac{x_{k+1}-x_k}{\varepsilon}\right)\left(\frac{x_k-x_{k-1}}{\varepsilon}\right). \tag{53}$$

That is, the easiest way to produce observable functionals involving powers of the velocities is to replace these powers by a product of velocities, each factor of which is taken at a slightly different time.

10. The Hamiltonian: Momentum

The Hamiltonian operator is of central importance in the usual formulation of quantum mechanics. We shall study in this section the functional corresponding to this operator. We could immediately define the Hamiltonian functional by adding the kinetic energy functional (52) or (53) to the potential energy. This method is artificial and does not exhibit the important relationship of the Hamiltonian to time. We shall define the Hamiltonian functional by the changes made in a state when it is displaced in time.

To do this we shall have to digress a moment to point out that the subdivision of time into *equal* intervals is not necessary. Clearly, any subdivision into instants t_i will be satisfactory; the limits are to be taken as the largest spacing, $t_{i+1} - t_i$, approaches zero. The total action S must now be represented as a sum

$$S = \sum_i S(x_{i+1}, t_{i+1}; x_i, t_i)\,, \tag{54}$$

where

$$S(x_{i+1}, t_{i+1}; x_i, t_i) = \int_{t_i}^{t_{i+1}} L(\dot{x}(t), x(t))dt\,, \tag{55}$$

the integral being taken along the classical path between x_i at t_i and x_{i+1} at t_{i+1}. For the simple one-dimensional example this becomes, with sufficient accuracy,

$$S(x_{i+1}, t_{i+1}; x_i, t_i) = \left\{\frac{m}{2}\left(\frac{x_{i+1} - x_i}{t_{i+1} - t_i}\right)^2 - V(x_{i+1})\right\}(t_{i+1} - t_i)\,; \tag{56}$$

the corresponding normalization constant for integration on dx_i is $A = (2\pi\hbar i(t_{i+1} - t_i)/m)^{-\frac{1}{2}}$.

The relation of H to the change in a state with displacement in time can now be studied. Consider a state $\psi(t)$ defined by a space-time region R'. Now imagine that we consider another state at time t, $\psi_\delta(t)$, defined by another region R'_δ. Suppose the region R'_δ is exactly the same as R' except that it is earlier by a time δ, i.e., displaced bodily toward the past by a time δ. All the apparatus to prepare the system for R'_δ is identical to that for R' but is operated a time δ sooner. If L depends explicitly on time, it, too, is to be displaced, i.e., the state ψ_δ is obtained from the L used for state ψ except that the time t in L_δ is replaced by $t + \delta$. We ask how does the state ψ_δ differe from ψ? In any measurement the chance of finding the system in a fixed region R'' is different for R' and R'_δ. Consider the change in the transition element $\langle\chi|1|\psi_\delta\rangle_{S_\delta}$ produced by the shift δ. We can consider this shift as effected by decreasing all values of t_i by δ for $i \leq k$ and leaving all t_i fixed for $i > k$, where the time t

lies in the interval between t_{k+1} and t_k.[19] This change will have no effect on $S(x_{i+1}, t_{i+1}; x_i, t_i)$ as defined by (55) as long as both t_{i+1} and t_i are changed by the same amount. On the other hand, $S(x_{k+1}, t_{k+1}; x_k, t_k)$ is changed to $S(x_{k+1}, t_{k+1}; x_k, t_k - \delta)$. The constant $1/A$ for the integration on dx_k is also altered to $(2\pi\hbar i(t_{k+1} - t_k + \delta)/m)^{-\frac{1}{2}}$. The effect of these changes on the transition element is given to the first order in δ by

$$\langle\chi|1|\psi\rangle_S - \langle\chi|1|\psi_\delta\rangle_{S_\delta} = \frac{i\delta}{\hbar}\langle\chi|H_k|\psi\rangle_S\,, \tag{57}$$

here the Hamiltonian functional H_k is defined by

$$H_k = \frac{\partial S(x_{k+1}, t_{k+1}; x_k, t_k)}{\partial t_k} + \frac{\hbar}{2i(t_{k+1} - t_k)}\,. \tag{58}$$

The last term is due to the change in $1/A$ and serves to keep H_k finite as $\varepsilon \to 0$. For example, for the expression (56) this becomes

$$H_k = \frac{m}{2}\left(\frac{x_{k+1} - x_k}{t_{k+1} - t_k}\right)^2 + \frac{\hbar}{2i(t_{k+1} - t_k)} + V(x_{k+1})\,,$$

which is just the sum of the kinetic energy functional (52) and that of the potential energy $V(x_{k+1})$.

The wave function $\psi_\delta(x, t)$ represents, of course, the same state as $\psi(x, t)$ will be after time δ, i.e., $\psi(x, t + \delta)$. Hence, (57) is intimately related to the operator equation (31).

One could also consider changes occasioned by a time shift in the final state χ. Of course, nothing new results in this way for it is only the relative shift of χ and ψ which counts. One obtains an alternative expression

$$H_k = -\frac{\partial S(x_{k+1}, t_{k+1}; x_k, t_k)}{\partial t_{k+1}} + \frac{\hbar}{2i(t_{k+1} - t_k)}\,. \tag{59}$$

This differs from (58) only by terms of order ε.

The time rate of change of a functional can be computed by considering the effect of shifting both initial and final state together. This has the same

[19] From the point of view of mathematical rigor, if δ is finite, as $\varepsilon \to 0$ one gets into difficulty in that, for example, the interval $t_{k+1} - t_k$ is kept finite. This can be straightened out by assuming δ to vary with time and to be turned on smoothly before $t = t_k$ and turned off smoothly after $t = t_k$. Then keeping the time variation of δ fixed, let $\varepsilon \to 0$. Then seek the first-order change as $\delta \to 0$. The result is essentially the same as that of the crude procedure used above.

effect as calculating the transition element of the functional referring to a later time. What results is the analog of the operator equation

$$\frac{\hbar}{i}\dot{\mathbf{f}} = \mathbf{Hf} - \mathbf{fH}\,.$$

The momentum functional p_k can be defined in an analagous way by considering the changes made by displacements of position:

$$\langle \chi|1|\psi\rangle_S - \langle \chi|1|\psi_\Delta\rangle_{S_\Delta} = \frac{i\Delta}{\hbar}\langle \chi|p_k|\psi\rangle_S\,.$$

The state ψ_Δ is prepared from a region R'_Δ which is identical to region R' except that it is moved a distance Δ in space. (The Lagrangian, if it depends explicitly on x, must be altered to $L_\Delta = L(\dot{x}, x - \Delta)$ for times previous to t.) One finds[20]

$$p_k = \frac{\partial S(x_{k+1}, x_k)}{\partial x_{k+1}} = -\frac{\partial S(x_{k+1}, x_k)}{\partial x_k}\,. \tag{60}$$

Since $\psi_\Delta(x,t)$ is equal to $\psi(x - \Delta, t)$, the close connection between p_k and the x-derivative of the wave function is established.

Angular momentum operators are related in an analogous way to rotations.

The derivative with respect to t_{i+1} of $S(x_{i+1}, t_{i+1}; x_i, t_i)$ appears in the definition of H_i. The derivative with respect to x_{i+1} defines p_i. But the derivative with respect to t_{i+1} of $S(x_{i+1}, t_{i+1}; x_i, t_i)$ is related to the derivative with respect to x_{i+1}, for the function $S(x_{i+1}, t_{i+1}; x_i, t_i)$ defined by (55) satisfies the Hamilton–Jacobi equation. Thus, the Hamilton–Jacobi equation is an equation expressing H_i in terms of the p_i. In other words, it expresses the fact that time displacements of states are related to space displacements of the same states. This idea leads directly to a derivation of the Schroedinger equation which is far more elegant than the one exhibited in deriving Eq. (30).

---

20 We did not immediately substitute p_i from (60) into (47) because (47) would then no longer have been valid to both zero order and the first order in ε. We could derive the commutation relations, but not the equations of motion. The two expressions in (60) represent the momenta at each end of the interval t_i to t_{i+1}. They differ by $\varepsilon V'(x_{k+1})$ because of the force acting during the time ε.

11. Inadequacies of the Formulation

The formulation given here suffers from a serious drawback. The mathematical concepts needed are new. At present, it requires an unnatural and cumbersome subdivision of the time interval to make the meaning of the equations clear. Considerable improvement can be made through the use of the notation and concepts of the mathematics of functionals. However, it was thought best to avoid this in a first presentation. One needs, in addition, an appropriate measure for the space of the argument functions $x(t)$ of the functionals.[10]

It is also incomplete from the physical standpoint. One of the most important characteristics of quantum mechanics is its invariance under unitary transformations. These correspond to the canonical transformations of classical mechanics. Of course, the present formulation, being equivalent to ordinary formulations, can be mathematically demonstrated to be invariant under these transformations. However, it has not been formulated in such a way that it is *physically* obvious that it is invariant. This incompleteness shows itself in a definite way. No direct procedure has been outlined to describe measurements of quantities other than position. Measurements of momentum, for example, of one particle, can be defined in terms of measurements of positions of other particles. The result of the analysis of such a situation does show the connection of momentum measurements to the Fourier transform of the wave function. But this is a rather roundabout method to obtain such an important physical result. It is to be expected that the postulates can be generalized by the replacement of the idea of "paths in a region of space-time R" to "paths of class R," or "paths having property R." But which properties correspond to which physical measurements has not been formulated in a general way.

12. A Possible Generalization

The formulation suggests an obvious generalization. There are interesting classical problems which satisfy a principle of least action but for which the action cannot be written as an integral of a function of positions and velocities. The action may involve accelerations, for example. Or, again, if interactions are not instantaneous, it may involve the product of coordinates at two different times, such as $\int x(t)x(t+T)dt$. The action, then, cannot be broken up into a sum of small contributions as in (10). As a consequence, no wave function is available to describe a state. Nevertheless, a transition probability can be defined for getting from a region R' into another R''. Most of the theory of the transition elements $\langle \chi_{t''}|F|\psi_{t'}\rangle_S$ can be carried over. One simply invents a symbol, such as $\langle R''|F|R'\rangle_S$ by an equation such as (39) but with the expressions (19) and (20) for ψ and χ substituted, and the more general action substituted for S. Hamiltonian and momentum functionals can be defined as in Sec. 10. Further details may be found in a thesis by the author.[21]

13. Application to Eliminate Field Oscillators

One characteristic of the present formulation is that it can give one a sort of bird's-eye view of the space-time relationships in a given situation. Before the integrations to the x_i are performed in an expression such as (39) one has a sort of format into which various F functionals may be inserted. One can study how what goes on in the quantum-mechanical system at different times is interrelated. To make these vague remarks somewhat more definite, we discuss an example.

In classical electrodynamics the fields describing, for instance, the interaction of two particles can be represented as a set of oscillators. The equations of motion of these oscillators may be solved and the oscillators essentially eliminated (Lienard and Wiechert potentials). The interactions which result involve relationships of the motion of one particle at one time, and of the other particle at another time. In quantum electrodynamics the field is again represented as a set of oscillators. But the motion of the

[21] The theory of electromagnetism described by J. A. Wheeler and R. P. Feynman, *Rev. Mod. Phys.* **17**, 157 (1945) can be expressed in a principle of least action involving the coordinates of particles alone. It was an attempt to quantize this theory, without reference to the fields, which led the author to study the formulation of quantum mechanics given here. The extension of the ideas to cover the case of more general action functions was developed in his Ph.D. thesis, "The principle of least action in quantum mechanics" submitted to Princeton University, 1942.

oscillators cannot be worked out and the oscillators eliminated. It is true that the oscillators representing longitudinal waves may be eliminated. The result is instantaneous electrostatic interaction. The electrostatic elimination is very instructive as it shows up the difficulty of self-interaction very distinctly. In fact, it shows it up so clearly that there is no ambiguity in deciding what term is incorrect and should be omitted. This entire process is not relativistically invariant, nor is the omitted term. It would seem to be very desirable if the oscillators, representing transverse waves, could also be eliminated. This presents an almost insurmountable problem in the conventional quantum mechanics. We expect that the motion of a particle a at one time depends upon the motion of b at a previous time, and *vice versa*. A wave function $\psi(x_a, x_b; t)$, however, can only describe the behavior of both particles at one time. There is no way to keep track of what b did in the past in order to determine the behavior of a. The only way is to specify the state of the set of oscillators at t, which serve to "remember" what b (and a) had been doing.

The present formulation permits the solution of the motion of all the oscillators and their complete elimination from the equations describing the particles. This is easily done. One must simply solve for the motion of the oscillators before one integrates over the various variables x_i for the particles. It is the integration over x_i which tries to condense the past history into a single state function. This we wish to avoid. Of course, the result depends upon the initial and final states of the oscillator. If they are specified, the result is an equation for $\langle \chi_{t''} | 1 | \psi_{t'} \rangle$ like (38), but containing as a factor, besides $\exp(iS/\hbar)$ another functional G depending only on the coordinates describing the paths of the particles.

We illustrate briefly how this is done in a very simple case. Suppose a particle, coordinate $x(t)$, Lagrangian $L(\dot{x}, x)$ interacts with an oscillator, coordinate $q(t)$, Lagrangian $\frac{1}{2}(\dot{q}^2 - \omega^2 q^2)$, through a term $\gamma(x, t)q(t)$ in the Lagrangian for the system. Here $\gamma(x, t)$ is any function of the coordinate $x(t)$ of the particle and the time.[22] Suppose we desire the probability of a transition from a state at time t', in which the particle's wave function is $\psi_{t'}$ and the oscillator is in energy level n, to a state at t'' with the particle

[22] The generalization to the case that γ depends on the velocity, $\dot{x}$, of the particle presents no problem.

in $x_{t''}$ and oscillator in level m. This is the square of

$$\langle \chi_{i''}\varphi_m|1|\psi_{t'}\varphi_n\rangle_{S_p+S_0+S_I} = \int\cdots\int \varphi_m^*(q_j)\chi_{t''}^*(x_j) \times \exp\frac{i}{\hbar}(S_p+S_0+S_I)\psi_{t'}(x_0)\varphi_n(q_0) \cdot\frac{dx_0}{A}\frac{dq_0}{a}\cdots\frac{dx_{j-1}}{A}\frac{dq_{j-1}}{a}dx_j dq_j\,. \qquad (61)$$

Here $\varphi_n(q)$ is the wave function for the oscillator in state n, S_p is the action

$$\sum_{i=0}^{j-1} S_p(x_{i+1},x_i)\,,$$

calculated for the particle as though the oscillator were absent,

$$S_0 = \sum_{i=0}^{j-1}\left[\frac{\varepsilon}{2}\left(\frac{q_{i+1}-q_i}{\varepsilon}\right)^2 - \frac{\varepsilon\omega^2}{2}q_{i+1}^2\right]$$

that of the oscillator alone, and

$$S_I = \sum_{i=0}^{j-1}\gamma_i q_i$$

(where $\gamma_i = \gamma(x_i,t_i)$) is the action of interaction between the particle and the oscillator. The normalizing constant, a, for the oscillator is $(2\pi\varepsilon i/\hbar)^{-\frac{1}{2}}$. Now the exponential depends quadratically upon all the q_i. Hence, the integrations over all the variables q_i for $0 < i < j$ can easily be performed. One is integrating a sequence of Gaussian integrals.

The result of these integrations is, writing $T = t'' - t'$, $(2\pi i\hbar\sin\omega T/\omega)^{-\frac{1}{2}}\exp i(S_p + Q(q_j,q_0))/\hbar$, where $Q(q_j,q_0)$ turns out to be just the classical action for the forced harmonic oscillator (see reference 15)†. Explicitly it is

$$Q(q_j,q_0) = \frac{\omega}{2\sin\omega T}\Bigg[(\cos\omega T)(q_j^2+q_0^2) - 2q_jq_0 + \frac{2q_0}{\omega}\int_{t'}^{t''}\gamma(t)\sin\omega(t-t')dt + \frac{2q_j}{\omega}\int_{t'}^{t''}\gamma(t)\sin\omega(t''-t)dt - \frac{2}{\omega^2}\int_{t'}^{t''}\int_{t'}^{t}\gamma(t)\gamma(s)\sin\omega(t''-t)\sin\omega(s-t')ds dt\Bigg]\,.$$

† Editor's note: This should read (see reference 21), i.e., Feynman's doctoral thesis.

It has been written as though $\gamma(t)$ were a continuous function of time. The integrals really should be split into Riemann sums and the quantity $\gamma(x_i, t_i)$ substituted for $\gamma(t_i)$. Thus, Q depends on the coordinates of the particle at all times through the $\gamma(x_i, t_i)$ and on that of the oscillator at times t' and t'' only. Thus, the quantity (61) becomes

$$\begin{aligned}&\langle\chi_{t''}\varphi_m|1|\psi_{t'}\varphi_n\rangle_{S_p+S_0+S_I}\\&\quad=\int\cdots\int\chi^*_{t''}(x_j)G_{mn}\exp\left(\frac{iS_p}{\hbar}\right)\psi_{t'}(x_0)\frac{dx_0}{A}\cdots\frac{dx_{j-1}}{A}dx_j\\&\quad=\langle\chi_{t''}|G_{mn}|\psi_{t'}\rangle_{S_p}\end{aligned}$$

which now contains the coordinates of the particle only, the quantity G_{mn} being given by

$$G_{mn}=(2\pi i\hbar\sin\omega T/\omega)^{-\frac{1}{2}}\iint\varphi^*_m(q_j)\exp(iQ(q_j,q_0)/\hbar)\varphi_n(q_0)dq_jdq_0\,.$$

Proceeding in an analogous manner one finds that all of the oscillators of the electromagnetic field can be eliminated from a description of the motion of the charges.

14. Statistical Mechanics: Spin and Relativity

Problems in the theory of measurement and statistical quantum mechanics are often simplified when set up from the point of view described here. For example, the influence of a perturbing measuring instrument can be integrated out in principle as we did in detail for the oscillator. The statistical density matrix has a fairly obvious and useful generalization. It results from considering the square of (38). It is an expression similar to (38) but containing integrations over two sets of variables dx_i and dx'_i. The exponential is replaced by $\exp i(S-S')/\hbar$, where S' is the same function of the x'_i as S is of x_i. It is required, for example, to describe the result of the elimination of the field oscillators where, say, the final state of the oscillators is unspecified and one desires only the sum over all final states m.

Spin may be included in a formal way. The Pauli spin equation can be obtained in this way: One replaces the vector potential interaction term in $S(x_{i+1}, x_i)$,

$$\frac{e}{2c}(\mathbf{x}_{i+1} - \mathbf{x}_i) \cdot \mathbf{A}(\mathbf{x}_i) + \frac{e}{2c}(\mathbf{x}_{i+1} - \mathbf{x}_i) \cdot \mathbf{A}(\mathbf{x}_{i+1})$$

arising from expression (13) by the expression

$$\frac{e}{2c}(\boldsymbol{\sigma} \cdot (\mathbf{x}_{i+1} - \mathbf{x}_i))(\boldsymbol{\sigma} \cdot \mathbf{A}(\mathbf{x}_i)) + \frac{e}{2c}(\boldsymbol{\sigma} \cdot \mathbf{A}(\mathbf{x}_{i+1}))(\boldsymbol{\sigma} \cdot (\mathbf{x}_{i+1} - \mathbf{x}_i)) .$$

Here $\mathbf{A}$ is the vector potential, $\mathbf{x}_{i+1}$ and $\mathbf{x}_i$ the vector positions of a particle at times t_{i+1} and t_i and $\boldsymbol{\sigma}$ is Pauli's spin vector matrix. The quantity Φ must now be expressed as $\prod_i \exp iS(x_{i+1}, x_i)/h$ for this differs from the exponential of the sum of $S(x_{i+1}, x_i)$. Thus, Φ is now a spin matrix.

The Klein Gordon relativistic equation can also be obtained formally by adding a fourth coordinate to specify a path. One considers a "path" as being specified by four functions $x^{(\mu)}(\tau)$ of a parameter τ. The parameter τ now goes in steps ε as the variable t went previously. The quantities $x^{(1)}(t)$, $x^{(2)}(t)$, $x^{(3)}(t)$ are the space coordinates of a particle and $x^{(4)}(t)$ is a corresponding time. The Lagrangian used is

$$\sum_{\mu=1}^{4}{}'[(dx^\mu/d\tau)^2 + (e/c)(dx^\mu/d\tau)A_\mu] ,$$

where A_μ is the 4-vector potential and the terms in the sum of $\mu = 1, 2, 3$ are taken with reversed sign. If one seeks a wave function which depends upon π periodically, one can show this must satisfy the Klein Gordon equation. The Dirac equation results from a modification of the Lagrangian used for the Klein Gordon equation, which is analagous to the modification of the non-relativistic Lagrangian required for the Pauli equation. What results directly is the square of the usual Dirac operator.

These results for spin and relativity are purely formal and add nothing to the understanding of these equations. There are other ways of obtaining the Dirac equation which offer some promise of giving a clearer physical interpretation to that important and beautiful equation.

The author sincerely appreciates the helpful advice of Professor and Mrs. H. C. Corben and of Professor H. A. Bethe. He wishes to thank Professor J. A. Wheeler for very many discussions during the early states of the work.

THE LAGRANGIAN IN QUANTUM MECHANICS*

P. A. M. DIRAC

Received 19 November 1932

Abstract

Quantum mechanics was built up on a foundation of analogy with the Hamiltonian theory of classical mechanics. This is because the classical notion of canonical coordinates and momenta was found to be one with a very simple quantum analogue, as a result of which the whole of the classical Hamiltonian theory, which is just a structure built up on this notion, could be taken over in all its details into quantum mechanics.

Now there is an alternative formulation for classical dynamics, provided by the Lagrangian. This requires one to work in terms of coordinates and velocities instead of coordinates and momenta. The two formulations are, of course, closely related, but there are reasons for believing that the Lagrangian one is the more fundamental.

In the first place the Lagrangian method allows one to collect together all the equations of motion and express them as the stationary property of a certain action function. (This action function is just the time-integral of the Lagrangian.) There is no corresponding action principle in terms of the coordinates and momenta of the Hamiltonian theory. Secondly the Lagrangian method can easily be expressed relativistically, on account of the action function being a relativistic invariant; while the Hamiltonian method is essentially non-relativistic in form, since it marks out a particular time variable as the canonical conjugate of the Hamiltonian function.

For these reasons it would seem desirable to take up the question of what corresponds in the quantum theory to the Lagrangian method of the classical theory. A little consideration shows, how-

* This article was first published in *Physikalische Zeitschrift der Sowjetunion*, Band 3, Heft 1 (1933), pp. 64–72.

ever, that one cannot expect to be able to take over the classical Lagrangian equations in any very direct way. These equations involve partial derivatives of the Lagrangian with respect to the coordinates and velocities and no meaning can be given to such derivatives in quantum mechanics. The only differentiation process that can be carried out with respect to the dynamical variables of quantum mechanics is that of forming Poisson brackets and this process leads to the Hamiltonian theory.[1]

We must therefore seek our quantum Lagrangian theory in an indirect way. We must try to take over the ideas of the classical Lagrangian theory, not the equations of the classical Lagrangian theory.

Contact Transformations

Lagrangian theory is closely connected with the theory of contact transformations. We shall therefore begin with a discussion of the analogy between classical and quantum contact transformations. Let the two sets of variables be p_r, q_r and P_r, Q_r, $(r = 1, 2, \ldots, n)$ and suppose the q's and Q's to be all independent, so that any function of the dynamical variables can be expressed in terms of them. It is well known that in the classical theory the transformation equations for this case can be put in the form

$$p_r = \frac{\partial S}{\partial q_r}, \qquad P_r = -\frac{\partial S}{\partial Q r}, \tag{1}$$

where S is some function of the q's and Q's.

In the quantum theory we may take a representation in which the q's are diagonal, and a second representation in which the Q's are diagonal.

[1] Processes of partial differentiation with respect to matrices have been given by Born, Heisenberg and Jordan (*ZS. f. Physik* **35**, 561, 1926) but these processes do not give us means of differentiation with respect to dynamical variables, since they are not independent of the representation chosen. As an example of the difficulties involved in differentiation with respect to quantum dynamical variables, consider the three components of an angular momentum, satisfying

$$m_x m_y - m_y m_x = ih m_z \,.$$

We have here m_z expressed explicitly as a function of m_x and m_y, but we can give no meaning to its partial derivative with respect to m_x or m_y.

There will be a transformation function $(q'|Q')$ connecting the two representations. We shall now show that this transformation function is the quantum analogue of $e^{iS/h}$.

If α is any function of the dynamical variables in the quantum theory, it will have a "mixed" representative $(q'|\alpha|Q')$, which may be defined in terms of either of the usual representatives $(q'|\alpha|q'')$, $(Q'|\alpha|Q'')$ by

$$(q'|\alpha|Q') = \int (q'|\alpha|q'')dq''(q''|Q') = \int (q'|Q'')dQ''(Q''|\alpha|Q')\,.$$

From the first of these definitions we obtain

$$(q'|q_r|Q'|) = q'_r(q'|Q') \tag{2}$$

$$(q'|p_r|Q'|) = -ih\frac{\partial}{\partial q'_r}(q'|Q') \tag{3}$$

and from the second

$$(q'|Q_r|Q'|) = Q'_r(q'|Q') \tag{4}$$

$$(q'|P_r|Q'|) = ih\frac{\partial}{\partial Q'_r}(q'|Q')\,. \tag{5}$$

Note the difference in sign in (3) and (5).

Equations (2) and (4) may be generalised as follows. Let $f(q)$ be any function of the q's and $g(Q)$ any function of the Q's. Then

$$\begin{aligned}(q'|f(q)g(Q)|Q') &= \iint (q'|f(q)|q'')dq''(q''|Q'')dQ''(Q''|g(Q)|Q')\\ &= f(q')g(Q')(q'|Q')\,.\end{aligned}$$

Further, if $f_k(q)$ and $g_k(Q)$, $(k = 1, 2, \ldots, m)$ denote two sets of functions of the q's and Q's respectively,

$$(q'|\Sigma_k f_k(q)g_k(Q)|Q') = \Sigma_k f_k(q')g_k(Q')\cdot(q'|Q')\,.$$

Thus if α is any function of the dynamical variables and we suppose it to be expressed as a function $\alpha(qQ)$ of the q's and Q's in a "well-ordered" way, that is, so that it consists of a sum of terms of the form $f(q)g(Q)$, we shall have

$$(q'|\alpha(qQ)|Q') = \alpha(q'Q')(q'|Q')\,. \tag{6}$$

This is a rather remarkable equation, giving us a connection between $\alpha(qQ)$, which is a function of operators, and $\alpha(q'Q')$, which is a function of numerical variables.

Let us apply this result for $\alpha = p_r$. Putting

$$(q'|Q') = e^{iU/h}\,, \tag{7}$$

where U is a new function of the q''s and Q''s we get from (3)

$$(q'|p_r|Q') = \frac{\partial U(q'Q')}{\partial q'_r}(q'|Q')\,.$$

By comparing this with (6), we obtain

$$p_r = \frac{\partial U(qQ)}{\partial q_r}$$

as an equation between operators or dynamical variables, which holds provided $\partial U/\partial q_r$ is well-ordered. Similarly, by applying the results (6) for $\alpha = Pr$ and using (5), we get

$$P_r = -\frac{\partial U(qQ)}{\partial Q_r}\,,$$

provided $\partial U/\partial Q_r$ is well-ordered. These equations are of the same form as (1) and show that the U defined by (7) is the analogue of the classical function S, which is what we had to prove.

Incidentally, we have obtained another theorem at the same time, namely that Eqs. (1) hold also in the quantum theory provided the right-hand sides are suitably interpreted, the variables being treated classically for the purpose of the differentiations and the derivatives being then well-ordered. This theorem has been previously proved by Jordan by a different method.[2]

The Lagrangian and the Action Principle

The equations of motion of the classical theory cause the dynamical variables to vary in such a way that their values q_t, p_t at any time t are connected with their values q_T, p_T at any other time T by a contact

[2] Jordan, *ZS. f. Phys.* 38, 513, 1926.

transformation, which may be put into the form (1) with q, $p = q_t$, p_t; Q, $P = q_T$, p_T and S equal to the time integral of the Lagrangian over the range T to t. In the quantum theory the q_t, p_t will still be connected with the q_T, p_T by a contact transformation and there will be a transformation function $(q_t|q_T)$ connecting the two representations in which the q_t and the q_T are diagonal respectively. The work of the preceding section now shows that

$$(q_t|q_T) \text{ corresponds to } \exp\left[i\int_T^t Ldt/h\right], \tag{8}$$

where L is the Lagrangian. If we take T to differ only infinitely little from t, we get the result

$$(q_{t+dt}|q_t) \text{ corresponds to } \exp\left[iL\,dt/h\right]. \tag{9}$$

The transformation functions in (8) and (9) are very fundamental things in the quantum theory and it is satisfactory to find that they have their classical analogues, expressible simply in terms of the Lagrangian. We have here the natural extension of the well-known result that the phase of the wave function corresponds to Hamilton's principle function in classical theory. The analogy (9) suggests that we ought to consider the classical Lagrangian, not as a function of the coordinates and velocities, but rather as a function of the coordinates at time t and the coordinates at time $t+dt$.

For simplicity in the further discussion in this section we shall take the case of a single degree of freedom, although the argument applies also to the general case. We shall use the notation

$$\exp\left[i\int_T^t L\,dt/h\right] = A(tT),$$

so that $A(tT)$ is the classical analogue of $(q_t|q_T)$.

Suppose we divide up the time interval $T \to t$ into a large number of small sections $T \to t_1$, $t_1 \to t_2, \ldots, t_{m-1} \to t_m$, $t_m \to t$ by the introduction of a sequence of intermediate times $t_1, t_2, \ldots, t_m$. Then

$$A(tT) = A(tt_m)A(t_m t_{m-1})\ldots A(t_2 t_1)A(t_1 T). \tag{10}$$

Now in the quantum theory we have

$$(q_t|q_T) = \int (q_t|q_m)dq_m(q_m|q_{m-1})dq_{m-1}\ldots(q_2|q_1)dq_1(q_1|q_T), \tag{11}$$

where q_k denotes q at the intermediate time t_k, $(k = 1, 2, \ldots, m)$. Equation (11) at first sight does not seem to correspond properly to Eq. (10), since on the right-hand side of (11) we must integrate after doing the multiplication while on the right-hand side of (10) there is no integration.

Let us examine this discrepancy by seeing what becomes of (11) when we regard t as extremely small. From the results (8) and (9) we see that the integrand in (11) must be of the form $e^{iF/h}$ where F is a function of $q_T, q_1, q_2 \ldots, q_m, q_t$ which remains finite as h tends to zero. Let us now picture one of the intermediate q's, say q_k, as varying continuously while the others are fixed. Owing to the smallness of h, we shall then in general have F/h varying extremely rapidly. This means that $e^{iF/h}$ will vary periodically with a vary high frequency about the value zero, as a result of which its integral will be practically zero. The only important part in the domain of integration of q_k is thus that for which a comparatively large variation in q_k produces only a very small variation in F. This part is the neighbourhood of a point for which F is stationary with respect to small variations in q_k.

We can apply this argument to each of the variables of integration in the right-hand side of (11) and otbain the result that the only important part in the domain of integration is that for which F is stationary for small variations in all the intermediate q's. But, by applying (8) to each of the small time sections, we see that F has for its classical analogue

$$\int_{t_m}^{t} L\,dt + \int_{t_{m-1}}^{t_m} L\,dt + \cdots + \int_{t_1}^{t_2} L\,dt + \int_{T}^{t_1} L\,dt = \int_{T}^{t} L\,dt\,,$$

which is just the action function which classical mechanics requires to be stationary for small variations in all the intermediate q's. This shows the way in which Eq. (11) goes over into classical results when h becomes extremely small.

We now return to the general case when h is not small. We see that, for comparison with the quantum theory, Eq. (10) must be interpreted in the following way. Each of the quantities A must be considered as a function of the q's at the two times to which it refers. The right-hand side is then a function, not only of q_T and q_t, but also of $q_1, q_2, \ldots, q_m$, and in order to get from it a function of q_T and q_t only, which we can equate to the left-hand side, we must substitute for $q_1, q_2, \ldots, q_m$ their values given by the

action principle. This process of substitution for the intermediate q's then corresponds to the process of integration over all values of these q's in (11).

Equation (11) contains the quantum analogue of the action principle, as may be seen more explicitly from the following argument. From Eq. (11) we can extract the statement (a rather trivial one) that, if we take specified values for q_T and q_t, then the importance of our considering any set of values for the intermediate q's is determined by the importance of this set of values in the integration on the right-hand side of (11). If we now make h tend to zero, this statement goes over into the classical statement that, if we take specified values for q_T and q_t, then the importance of our considering any set of values for the intermediate q's is zero unless these values make the action function stationary. This statement is one way of formulating the classical action principle.

Application to Field Dynamics

We may treat the problem of a vibrating medium in the classical theory by Lagrangian methods which form a natural generalisation of those for particles. We choose as our coordinates suitable field quantities or potentials. Each coordinate is then a function of the four space-time variables x, y, z, t, corresponding to the fact that in particle theory it is a function of just the one variable t. Thus the one independent variable t of particle theory is to be generalised to four independent variables x, y, z, t.[3]

We introduce at each point of space-time a Lagrangian density, which must be a function of the coordinates and their first derivatives with respect to x, y, z and t, corresponding to the Lagrangian in particle theory being a function of coordinates and velocities. The integral of the Lagrangian density over any (four-dimensional) region of space-time must then be stationary for all small variations of the coordinates inside the region, provided the coordinates on the boundary remain invariant.

[3] It is customary in field dynamics to regard the values of a field quantity for two different values of (x, y, z) but the same value of t as two different coordinates, instead of as two values of the same coordinate for two different points in the domain of independent variables, and in this way to keep to the idea of a single independent variable t. This point of view is necessary for the Hamiltonian treatment, but for the Lagrangian treatment the point of view adopted in the text seems preferable on account of its greater space-time symmetry.

It is now easy to see what the quantum analogue of all this must be. If S denotes the integral of the classical Lagrangian density over a particular region of space-time, we should expect there to be a quantum analogue of $e^{i\,S/h}$ corresponding to the $(q_t|q_T)$ of particle theory. This $(q_t|q_T)$ is a function of the values of the coordinates at the ends of the time interval to which it refers and so we should expect the quantum analogue of $e^{i\,S/h}$ to be a function (really a functional) of the values of the coordinates on the boundary of the space-time region. This quantum analogue will be a sort of "generalized transformation function". It cannot in general be interpreted, like $(q_t|q_T)$, as giving a transformation between one set of dynamical variables and another, but it is a four-dimensional generalization of $(q_t|q_T)$ in the following sense.

Corresponding to the composition law for $(q_t|q_T)$

$$(q_t|q_T) = \int (q_t|q_1)dq_1(q_1|q_T)\,, \tag{12}$$

the generalized transformation function (g.t.f.) will have the following composition law. Take a given region of space-time and divide it up into two parts. Then the g.t.f. for the whole region will equal the product of the g.t.f.'s for the two parts, integrated over all values for the coordinates on the common boundary of the two parts.

Repeated application of (12) gives us (11) and repeated application of the corresponding law for g.t.f.'s will enable us in a similar way to connect the g.t.f. for any region with the g.t.f.'s for the very small sub-regions into which that region may be divided. This connection will contain the quantum analogue of the action principle applied to fields.

The square of the modulus of the transformation function $(q_t|q_T)$ can be interpreted as the probability of an observation of the coordinates at the later time t giving the result q_t for a state for which an observation of the coordinates at the earlier time T is certain to give the result q_T. A corresponding meaning for the square of the modulus of the g.t.f. will exist only when the g.t.f. refers to a region of space-time bounded by two separate (three-dimensional) surfaces, each extending to infinity in the space directions and lying entirely outside any light-cone having its vertex on the surface. The square of the modulus of the g.t.f. then gives the probability of the coordinates having specified values at all points on the